Androids

Build Your Own Lifelike Robots

Bryan Bergeron, MD

Thomas B. Talbot, MD

New York   Chicago   San Francisco
Athens   London   Madrid
Mexico City   Milan   New Delhi
Singapore   Sydney   Toronto

McGraw-Hill Education books are available at special quantity discounts to use as premiums and sales promotions or for use in corporate training programs. To contact a representative, please visit the Contact Us page at www.mhprofessional.com.

Androids: Build Your Own Lifelike Robots

1 2 3 4 5 6 7 8 9 0   DOC/DOC   1 9 8 7 6 5 4 3

ISBN   978-0-07-181404-1
MHID   0-07-181404-3

This book is printed on acid-free paper.

Sponsoring Editor Roger Stewart	**Copy Editor** James Madru
Editing Supervisor Stephen M. Smith	**Proofreader** Claire Splan
Production Supervisor Pamela A. Pelton	**Indexer** Jack Lewis
Acquisitions Coordinator Amy Stonebraker	**Art Director, Cover** Jeff Weeks
Project Manager Patricia Wallenburg, TypeWriting	**Composition** TypeWriting

This book is dedicated to all the children who take things apart
and all those who grew up to invent and create.

Special thanks to Dianne Talbot and all parents who tolerated
their childrens' inventive behaviors.

Special thanks, too, to Adam Wyatt for his android photography.

About the Authors

Bryan Bergeron, MD is author of two dozen books and 500+ magazine articles, holds several patents in the areas of human simulation and embedded systems, and is Editor-in-Chief of *SERVO Magazine* and *Nuts & Volts Magazine*. He has designed and built human simulators for academia, the military, and commercial markets. His previous book for McGraw-Hill was *Teardowns: Learn How Electronics Work by Taking Them Apart* (2010).

Thomas B. Talbot, MD is the creator of several systems for training medics and healthcare professionals, and founded the medical simulation research and development program at the U.S. Defense Department. The author of numerous articles and book chapters, he is currently creating logical and emotionally expressive artificial intelligence–driven virtual humans for medical education.

Contents

Introduction

*A*ndroids: Build Your Own Lifelike Robots takes a unique, fun approach to learning about embedded systems, robotics, and electronics by using the human body as the focus of discussion. Sure, robotic crawlers and carpet roamers are cool platforms for experimentation, but most are cold, lifeless creations that are hard for people to relate to. Humans, in contrast, have complex reflexes, from hearts that beat at different rates based on stress to pupils that adjust to ambient light levels. If you want to take the first step in making your robots closer to humans, then this book is for you.

The inspiration for this book stems in part from the vision of great science fiction writers and in part from our experience developing human cybernetic systems for saving real human lives. If you've seen *Blade Runner*, *Battlestar Galactica*, *Alien*, *Terminator*, *Prometheus*, or *Doctor Who*, then you know that the pinnacle of robotics isn't simply a thinking tin can, regardless of intelligence. The androids featured in these and other sci-fi classics not only pass the Turing test, but they are also physiologically correct—they breathe, bleed, and sweat as we do. As a result, short of surgical exploration, these machines are indistinguishable from humans.

The second, more immediate and practical inspiration from this book reflects the need for lifelike human surrogates to train healthcare professionals on how to save lives and treat real patients. Our experience with available surrogates has taught us that we're simply not there yet. Most of these systems are little more than storefront mannequins with a few sensors. What's needed is a cohort of eager, enthusiastic engineers, experimentalists, and inventors to create the next generation of human surrogates. We hope that this book represents the first step in that journey for you. Even if your ultimate goal is to create an android companion worthy of *Blade Runner*, in the meantime, you'll need a challenging day job to pay the bills. Designing human cybernetic systems that can help to save lives is a great place to start.

In developing this book, we've made a few assumptions about you, the reader. We assume that you:

- Have read at least a couple introductory books on electronics.
- Have programmed one of the popular microcontrollers, especially the Arduino.

- Have some experience in basic robotics construction, either from a kit or from following a Web/magazine article.
- Are aware of—and practice—basic electrical safety precautions.
- Have and use eye protection. (If you happen to launch a piece of wire into your eye with diagonal cutters, your vision could be permanently degraded—at least until bionic implants are available.)
- Have a natural and insatiable inquisitiveness.

Goals

We have done our best to ensure that after reading this book, you will

- Have a better understanding of cybernetics as applied to human systems. Each chapter should give you mental anchor points for understanding the physiologic mechanisms behind each project.
- Better understand how to apply microcontrollers, sensors, and actuators in the modeling and simulation of human systems.
- Be able to apply embedded systems to human cybernetic systems not covered in this book. The human body is a complex machine—with thousands of interconnected systems.
- Better understand the practical considerations that go into designing and constructing a cybernetic system. It's possible to spend thousands of dollars on silicone props that are nearly indistinguishable from the real thing, but what really matters at this point in your journey is that you understand the underlying principles. We're not building a Number 6 Cylon from *Battlestar Galactica* here—that's the focus of a follow-up book.

Organization

Androids: Build Your Own Lifelike Robots is organized by system and arranged in order of increasing project complexity. Feel free to skip ahead to a project of interest, but consider at least skimming the preceding chapters. In some cases an earlier project will be repurposed to support a more advanced project.

Each chapter is organized along the following outline:

- **Biological basis**   The biological underpinnings for the system(s) discussed in the chapter.
- **Relevance to androids**   How the biology translates to androids.
- **Experiments**   Hands-on experiments with circuits, Arduinos, source code, and sensors.
- **Modifications**   How to get more out of the basic experiments.
- **Gremlins**   Common problems and issues.
- **Search terms**   Search terms for readers wanting to learn more.

Chapter 1: Reflex Arc

In Chapter 1, we'll explore the basics of reflexes with the help of a pair of servos, a few sensors, and an Arduino microcontroller. After implementing a series of circuits patterned after the basic biological unit of the reflex, the reflex arc, you'll have an arsenal of reflex templates that you can expand and apply to your android designs.

Chapter 2: Behavior Chains

Building on the basic reflex arc, Chapter 2 explores behavior chains and how they can be used to create cybernetic personalities. Using a 5-degree-of-freedom (DOF) robotic arm, heat and light sensors, and a few light-emitting diodes (LEDs), we'll illustrate how to create a robot with autonomous, predictable behavior.

Chapter 3: Homeostasis

In Chapter 3, we explore thermal homeostasis and basic control theory using an Arduino, temperature sensors, and heat sources. We'll leverage the proportional, integral, derivative (PID) library for the Arduino so that we can focus on the application of control theory as opposed to low-level implementation.

Chapter 4: Light and Vision

In Chapter 4, we explore light sensors that can give your android some degree of vision. We also create a pair of eyes that with variable pupils and gaze follows a light source. This will give your android the appearance of a living face for human interaction.

Chapter 5: All Ears

In Chapter 5, we explore sound localization and mimic the normal eye and head movements associated with sound tracking. We'll work with directional and omnidirectional microphone sensors, silicone rubber "ears," a pair of passive infrared (PIR) sensors, a servo-controlled turret, an Arduino, and a high-performance ChipKit Arduino clone.

Chapter 6: A Heartbeat Away

In Chapter 6, we'll explore the acoustic and tactile feedback we expect from a living human—heart and breath sounds. We'll use an MP3 shield, an Arduino, a few Hall effect sensors, and a surface transducer to create realistic heat and breathing sounds. We'll also explore how to make those sounds respond to behavior triggers in the internal and external environments.

Chapter 7: If It Bleeds, Can We Kill It?

In Chapter 7, you'll learn how to give your android a circulatory system with a lifelike pulse. We'll work with a fluid pump, pressure sensors, and a pulse-monitor shield for the Arduino to create pulsing pressure waves to provide appendages with lifelike pulses that respond to external and internal conditions. This chapter leverages the system developed in Chapter 3.

Chapter 8: Simply a Matter of Time

Ever notice how every otherwise perfect android is ultimately discovered because the creator neglected to simulate aging? Well, we'll address that common fault in this chapter by exploring how biological systems age and, more important, how to mimic the appropriate behaviors in cybernetic systems. We'll build on the systems developed in Chapters 2 and 6.

Chapter 9: Affect and Expression

In Chapter 9, we explore several affordable approaches to making an expressive android that is a joy to work with. We leverage an EMIC-2 text-to-speech module, a pair of 8 × 8 bicolor LED matrices, and a handful of multicolored LEDs to create an expressive android head that talks.

Resources

The resources section provides a list of resources for sensors, actuators, and Arduino-compatible hardware, as well as technical reference sites.

Downloadable Code

See www.mhprofessional.com/Androids for fully documented code.

A Word about Hardware

We've made a conscious effort to keep the projects simple and affordable. And although most of these projects assume an Arduino Uno or equivalent, feel free to work with an Arduino Leonardo, Mega, or 80-MHz ChipKit Uno32, depending on what you have on hand. Similarly, although we may list a particular brand of shield, sensor, or output device, don't feel compelled to buy the same if you have something equivalent in your inventory. For example, although we use a standard Lynxmotion robot arm in Chapter 2, you can get by with a servo and do-it-yourself (DIY) gripper.

The same goes for prototyping kits. If you're teaching a class and build time is critical, then you really should consider the various prototyping kits and devices available. We've featured the Grove platform in a few of the projects, but several equivalent or superior products are available.

Finally, we haven't received any support from a vendor in putting together these projects and have no financial interest in the suppliers or manufacturers mentioned. We purchased the products featured in this book at regular retail prices. The only exception is the android body featured in Chapter 9, which is borrowed courtesy of the U.S. Army's Telemedicine and Advanced Technology Research Center (TATRC).

Androids

CHAPTER 1

Reflex Arc

You have the time and ability to focus on this page in part because you have reflexes—semiautonomous movements that you're not consciously directing. And because most of your reflexes rely on neurons in your body and spinal chord, more of the neurons in your brain can be devoted to tasks such as reading as well as contemplating the answers to life, the universe, and everything. Androids and semiautonomous vehicles can and do benefit from reflexes as well to free up precious computational resources for higher tasks such as simultaneous localization and mapping (SLAM).

In this chapter we'll explore the basics of reflexes with the help of a pair of servos, a few sensors, and an Arduino microcontroller. After implementing a series of circuits patterned after the basic biological unit of the reflex, the *reflex arc*, you'll have an arsenal of reflex templates that you can expand and apply to your android designs.

Biological Basis

It can be difficult to appreciate the arsenal of reflexes you have at your disposal to respond to internal and environmental events because many reflexes aren't activated until something goes awry. To illustrate, consider the following scenario.

Monkey Business

Imagine that you're hiking among the eucalyptus trees in the Presidio in San Francisco, and suddenly, without warning, a 75-pound genetically enhanced monkey jumps on your back. If you're in decent physical shape, you may manage to stay upright, grab the beast by the neck, and throw it to the ground before it pokes you in the eyes and gnaws off your ears. If, on the other hand, you buckle at the knees and find yourself on your back, blind, and having a hard time breathing

1

because something has jumped on your chest, then things may not work out so well for you in the end.

Let's assume the former case and examine what happens in your body in the first few milliseconds of the attack—before you're even consciously aware that your legs and back are supporting an additional 75 pounds. As a result of the jolt, thousands of sensors throughout your body fire—in your joints, in the soles of your feet, in your muscles and tendons, and in your organs—resulting in thousands of reflexes. Some of these reflexes direct your leg muscles to absorb and dissipate the impact of the beast, enabling you to use your arms and upper body to fend off the monkey instead of having to consciously maintain balance.

The Reflex Arc

To better understand what's happening in your body, we can model your reflexes at different levels and with different levels of granularity from minute changes in the folding of proteins in your muscles to gross changes in your behavior. For now, let's consider how events unfold at the level of your nerves, muscles, and sensors. Specifically, let's focus on a single muscle spindle sensor—a stretch sensor— embedded in one of your quadriceps muscles (or quads) in your right thigh. This muscle spindle sensor fires when your quad muscle is stretched, and the rate of firing increases with the velocity of stretch.

When the monkey lands on your back, your quad is suddenly and significantly stretched, causing the stretch receptor to fire rapidly, sending a stream of electrochemical pulses down a nerve fiber to your spinal chord. There the signal propagates across a gap to a second nerve fiber that connects, via yet another synapse, to muscle fibers in your quad. As a result, your quad contracts, counteracting the weight of the monkey. This same stretch reflex occurs with dozens of spindle fibers and thousands of muscle fibers in each of your quads and in other muscles throughout your body.

This arc or pathway from a sensor embedded in the muscle to the spinal chord, across a gap, and back to muscle fibers is referred to as a *reflex arc*. The components of a stretch reflex, an example of a simple reflex arc, are shown schematically in Figure 1-1. Note the reflex arc is a one-way connection from sensor to sensory fiber, across a space or synapse, to a motor fiber, across a neuromuscular junction, and terminating in a group of muscle fibers. The signal from the muscle spindle sensor changes with the contraction of the muscle, thereby providing a feedback loop for the stretch reflex.

Of course, this scenario is a simplified account of the complex cascade of events that occur within your body during the hypothetical attack. In addition to contracting your quads, there is a reflex relaxation of the muscles that oppose your quads—the hamstrings, the muscles on the backside of your thighs—so that your quads don't have to work against your hamstrings to keep your leg extended. There is also a reflex, based on the stress on the tendons, that inhibits your quads

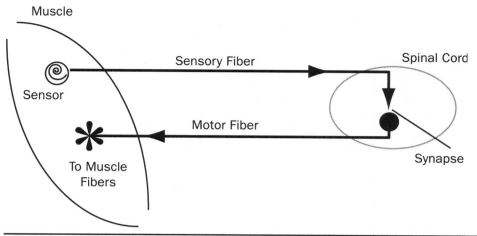

FIGURE 1-1   Components of the stretch reflex, an example of a simple reflex arc.

from contracting so rapidly and violently that they rip the tendons anchoring your muscles to your skeleton. The bottom line is that, thanks to thousands of reflex arcs throughout your body, you're able to die another day.

Sensors and their related reflex arcs have a threshold. Like digital electronic systems, once a sensor fires, it fires at full amplitude, with a pulse rate that reflects the degree of activation. For example, if you're attacked by a 6-ounce flying squirrel instead of a 75-pound monkey, then fewer spindle fiber sensors will be activated, and those that are activated fire at a lower pulse rate. As a result, fewer muscle fibers in your quads are activated less frequently, resulting in a less forceful contraction overall.

Another property of reflexes is that they have an *absolute refractory period*, or period of time after firing during which they can't be retriggered (Figure 1-2). Think of the absolute refractory period as the finite time required to reset the sensor.

Reflexes also exhibit a *relative refractory period*, the period following the absolute refractory period during which the sensor is relatively insensitive to triggering. As shown in Figure 1-2, insensitivity falls off exponentially after the absolute refractory period.

Pain and Suffering

Consider a second scenario in which you're at a workbench assembling a prototype leg for an android. As you reach for a linear actuator, you accidentally brush the back of your hand against the hot tip of a soldering iron. In response to the sudden, localized rise in skin temperature, thermal pain receptors (*nociceptors*) send streams of pulses along fibers to your spinal chord. Within milliseconds, signals travel through a reflex arc that includes a synapse in your spinal cord and a motor fiber to

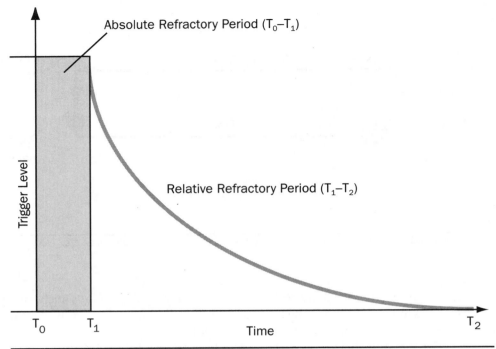

FIGURE 1-2   Absolute and relative reflex refractory periods.

muscle fibers in your arm and hand. As a result, your hand recoils from the soldering iron tip. If your reflexes are fast enough, you might end up with only a minor skin irritation and continue on with your work, not giving the incident a thought. On the other hand, if you had to rely on the smell of roasting skin to alert you to consciously move your hand, you'd have a serious burn, a possible infection, and a nasty scar.

The point of this scenario is that in addition to enabling you to maintain conscious focus on the task at hand, an advantage of a reflex action over deliberate action is speed. Conscious control, which may involve signals traversing hundreds or thousands of connections, or *neural synapses*, in your brain, is simply slower than a reflex arc involving a pair of local neurons. If you had to consciously blink every time a bug or dirt touched your eyelashes, you'd probably be blind by now.

Consider the stylized schematic of the reflex arc shown in Figure 1-3, in which there is a thermal pain receptor in the skin of your forearm, an inbound fiber from the sensor to your spinal cord, and a connection within your spinal cord to a pair of outbound motor fibers that terminate in muscle fibers in your arm and hand. There is also a third neural fiber from the brain that, when active, *inhibits* the reflex arc. Although this conscious inhibition doesn't have a place in the soldering-iron

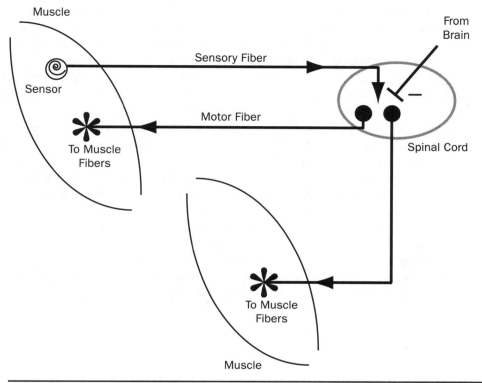

FIGURE 1-3 Schematic of a complex reflex arc with multiple synapses and an inhibitory connection from the brain.

scenario—unless you have some serious psychological issues—it would hopefully prevent you from, say, involuntarily releasing a too-hot cup of cocoa onto your lap.

Pain is an indicator that something is wrong and often that you need to take some action—or stop an action—to avoid injury. Some pain receptors respond as a function of stimulus intensity, whereas others switch from off to on as soon as some threshold is reached.

Triggers of pain receptors linked to reflex arcs include extremes of temperature, pressure, sound, and light. Because of the inhibitory connections from your brain, you may be able to override many of these reflex responses to pain—finishing a marathon with a sprint even though your muscles are aching, for example. You'll probably find some reflexes stubbornly resistant to conscious control—such as inhibiting your blink reflex in response to a sudden puff of air or flash of light.

Chemical Supercharger

Your nervous system is an electrochemical signal processor mediated by hundreds of different chemicals. When it comes to reflexes, one chemical worth exploring is adrenaline (epinephrine), which is pumped into your bloodstream by your adrenal glands when you're under stress. The action of adrenaline is complex and affects different tissues differently. However, one overall effect is to temporarily increase your muscular strength and endurance.

Adding to our stylized schematic of the reflex arc with multiple synapses and an inhibitory connection from your brain, we can model the overall effect of adrenaline as shown in Figure 1-4. Unlike electrochemical signals that travel on fibers from your brain or sensors, adrenaline doesn't generate a signal along a sensory or muscle fiber. What it does is effectively lower the threshold for sensor activation and maximize potential muscle contraction once the reflex arc is triggered. Using the metaphor of a vacuum tube or metal-oxide semiconductor field-effect transistor (MOSFET), adrenaline acts on the control grid or gate, changing the effective gain (or loss) of the system.

To explore the effect of adrenaline on the reflex arc, let's return to the monkey-on-your-back scenario. Imagine that as soon as you throw the monkey to the ground, you sprint 100 yards to a place that you feel safe from the primate. As you look back on the site of the attack, you realize that you covered the distance in a personal-best time of 12 seconds. Not bad, considering you're wearing hiking boots.

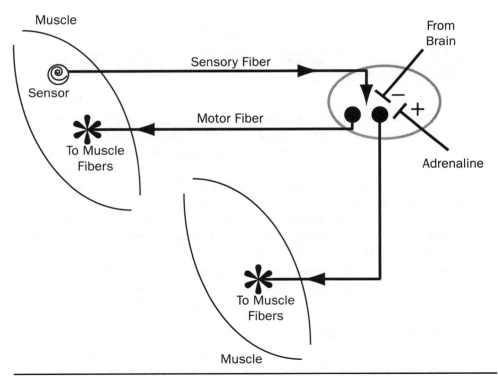

Figure 1-4 Schematic of a complex reflex arc with potentiating effects from adrenaline.

You were able to accomplish this feat because the adrenaline essentially supercharged your nervous system, with effects ranging from increasing your heart rate and pressure to upping the flow of blood and nutrients to your muscles. While temporarily useful, you wouldn't want to be jacked up on adrenaline long term. Not only would you expend energy more rapidly than normal, but you'd be at risk for stroke and heart attack. The same holds true for redlining an android—machines don't last long when they're operated at the edge of their design limits.

Personality

Perhaps you're saying to yourself, "Wait a minute, I wouldn't run from a monkey, genetically engineered or otherwise; I'd stand and fight to the death." And perhaps you would. The preference for fight versus flight reflects your overall personality. The boost from adrenaline can help one person run for the hills and give you the strength to stand and fight.

The point is that reflexes are generally consistent with personality. For example, if you're a trained soldier, then the appropriate reflex response to a flash of light or a loud sound is an offensive posture, ready for attack. Conversely, if you're a quiet, introverted librarian, then a more defensive posture would be more consistent with your personality. Of course, there are exceptions to the norm for personality and related reflex actions, and mental homes and jails are filled with examples of those exceptions. This is worth considering when you're designing your android's personality. We'll dive deeper into personality in Chapter 2.

Relevance to Android Designs

Given the advantage of human reflexes, it's obvious that androids can benefit from similar capabilities. For example, offloading the processing of sensors and effectors from the main processing unit to peripheral microcontrollers not only frees up computational resources for more complex tasks but also should lessen the chance of catastrophic single-point failure. However, realizing the potential benefits of reflexes requires that you carefully consider how they fit with your overall android designs.

A major design issue in enabling an android with reflexes is the degree to which an android with autonomous reflexes should be insulated from its environment. To put this issue in a practical context, imagine that your search-and-rescue android is in a partially collapsed mine shaft searching for survivors. Now let's say that the android's vibration sensors pick up high-intensity, low-frequency vibrations indicative of an imminent ceiling collapse. Should the android stay put and continue the search or run out of the mine shaft and head for cover? There's no one correct answer, of course, because it depends in part on how many other androids are available for the mission, the likelihood of finding survivors, the value (or net worth) of the survivors, and the relative value of the android. At $100, the android may be disposable, but at $10,000, it may be worth protecting—unless you happen to be the one trapped in the rubble.

Another major design issue is consistency of purpose with both responses and reflexes. For example, if your android is a battle bot, then the appropriate reflex response to a potential threat is an offensive posture, ready for attack. If, on the other hand, your android is designed as a domestic servant, then a defensive posture may be more appropriate. If you're out to build a depressed robot, like the depressed Marvin robot conceived by Douglas Adams, then reflexes should be slow and deliberate.

A low-level but critical design issue is how to handle multiple sensors that contribute to a single reflex action. One approach is to use *subsumption architecture*, a hierarchical scheme in which certain reflexes overrule or inhibit others. Following this architecture, you could manually or automatically assign relative rankings to reflex arcs. For example, let's say that you deem it more important for your android to avoid excessive temperatures than mechanical overload. Thermal sensors and their related reflex arcs can be designed to override any sensor inputs associated with servo load.

A limitation of the subsumption architecture is that it doesn't handle complexity well. Imagine an android with 50 reflex arcs and how you would define the reflex hierarchy. To use human physiology as a model, imagine the absurd case of a bee stinging you on your leg just as a monkey jumps on your back. Do you swat the bee first and then tend to the monkey or vice versa? Just as there's no one answer that works in every situation, even if you could think through the range of possible conditions and decide how your android should respond, it would be impractical to hard code such a hierarchy of reflex arcs on a mere microcontroller.

Another approach to handling potentially conflicting input from multiple sensors is to use *sensor fusion*, which is a complex way of saying sensor signal mixing. There are hundreds of approaches to sensor fusion, from simply averaging signals to applying complex digital signal-processing filters to the outputs of multiple different sensors. Sensor fusion has the potential to outperform subsumption in complex situations, such as when your android steps on a banana peel and starts to wobble. A properly tuned sensor-fusion algorithm can combine the signals from relevant sensors in a way that gives a meaningful, real-time answer to which reflexes should be fired and which should be inhibited.

With the theoretical underpinnings out of the way, let's get to work building some reflex simulators. We'll start with a basic biological reflex arc simulator using a standard servo, a few components, and an Arduino Uno microcontroller. When we're done, you'll have multiple circuits that you can apply directly to your android design, as well as the basis for more complex reflexes.

Simple Reflex Arc Experiments

Time to roll up your sleeves and apply theory to practice, starting with a simulation of the simple reflex arc simulator patterned after the simple reflex arc discussed earlier. The difference here is that the synapse in the spinal cord is implied, and the spindle fiber stretch sensor is represented first by a momentary contact switch and then with a force sensor. As you'll see, the simplicity of this simulator design lends itself to use in virtually any android design.

Bill of Materials

To construct the series of simple reflex arc simulators described in the first part of this chapter, you'll need the following:

- Analog hobby servos (2)
- Momentary contact switches (2)
- 10-kΩ potentiometers (2)
- Arduino Uno
- 5-V direct-current (dc) power supply
- 10-kΩ, ¼-W resistor
- Force-sensing resistor (1)
- Jumpers or wires
- Prototyping shield (optional)

Don't worry about brands or models at this point; just about any switch, analog servo, potentiometer, and Arduino-compatible microcontroller will do. For example, we use a mounted micro–push button for momentary contact switches, but you're free to use two bare wires, a reed switch and magnet trigger, or a phototransistor and light-emitting diode (LED) trigger. For potentiometers, you can experiment with an audio taper potentiometer if you like because it more closely resembles the motion about a joint with muscle contraction. However, because audio taper pots can be touchy when used with some servos, we prefer linear taper pots.

If you don't happen to have parts lying around and need to order something, you might consider some of the elements in our experimental setup, shown in Figure 1-5. We have a pair of analog hobby servos, a pair of momentary contact switches, a pair of 10-kΩ potentiometers, and a force-sensing resistor mounted on a terminal block (more on this later). The Arduino Uno microcontroller is hidden by a Grove base shield.

We're fans of solderless prototyping shields because they make setup and breakdown a breeze. The downside of the Grove shield shown in Figure 1-5 is the lack of a breadboard prototyping area to hang additional components. Another option is the screw shield available from Sparkfun and Maker Shed, which offers screw-terminal connections for each Arduino pin and a central prototyping area but lacks solderless jacks for external components. If you don't use a shield, at least consider using male-male jumpers to connect the servos to the Arduino. Packs of

Figure 1-5   Components of the authors' experimental setup.

colorful, easy-to-trace jumpers are available in male-male, male-female, and female-female configurations from Sparkfun and Maker Shed.

Although we've seen examples where a servo is powered by the Arduino's 5-V power supply, it's a bad idea. Either use a separate 5-V supply for the servo, or power the entire board with a regulated external 5-V supply. If you don't use a separate supply for the servos, you risk burning out the 5-V regulator on the Arduino. Consider that while the Arduino Uno is rated at 40 mA maximum per pin, a typical hobby servo, such as the HiTec HS-422, draws about 140 mA with no load. If you put the servo in a bind, current draw is even greater.

As an aside, our standard experimental setup is a pair of Arduinos mounted on a piece of particleboard with a shared power supply. We use one Arduino for production and the other to test code and hardware. In this way, we keep one Arduino with the latest working version of a project and use the other to untangle problems. This isn't necessary, but we've found it to be a real timesaver.

Circuit

The first circuit, that of a simple reflex arc simulator, is show in Figure 1-6. It consists of a standard analog hobby servo for the muscle, a momentary closed switch for

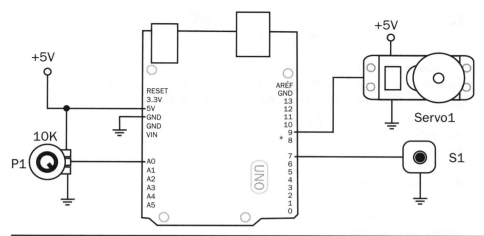

FIGURE 1-6   Schematic of bare-bones simple reflex arc simulator. P1 controls servo position, and S1 initiates the reflex.

the muscle spindle fiber, and a potentiometer to set the position of the muscle. The Arduino Uno is configured to simulate the signals produced by the muscle spindle fiber or stretch receptor as well as the motor signals to the muscle.

The 10-kΩ pot, P1, forms a voltage divider, with the wiper going to analog input A0. A momentary contact switch, S1, grounds digital input pin D7 when depressed. The servo, Servo1, receives pulse-width-modulated (PWM) output from digital pin D9. Note the 5-Vdc supply connected directly to the servo—and via the 5-Vdc onboard regulator.

Construction

Constructing this simulator is straightforward. Either fit the servo with a commercial plastic or aluminum bracket or simply place it on your desktop. If you have a bracket for your servo, then you have the option of mounting the switch on the servo horn assembly. And don't worry about the parts that aren't used—we'll get to those shortly in the following projects.

Code

The Arduino code for the simple reflex arc simulator, shown in Listing 1-1, leverages the standard servo library. This library hides the complexity of generating PWM signals to control the position and rate of movement of analog servos.

The listing is a modification of the servo example program that ships with the Arduino 1.0 programming environment. The key difference is that when button S1 is pressed, the position of the servo is saved, and then the reflex routine is run. Following the reflex, the servo returns to the position corresponding to the value of potentiometer P1.

LISTING 1-1 Arduino code for simple reflex arc simulator.

```
/*
Simple Reflex Arc Simulator
Androids: Build Your Own Lifelike Robots by Bergeron/Talbot
Listing 1-1. See www.mhprofessional.com/Androids for fully
documented code
Arduino 1.0.1 environment
*/

#include <Servo.h>
Servo myservo;
const int pot1pin = A0;
const int stretchButtonPin = 7;
const int servo1pin = 9;
int pot1val;
int servoDelay = 15;
int servoReflexDelay = 2;
int stretchButtonState = 0;
int servo1Position = 0;

void setup() {
  myservo.attach(servo1pin);
  pinMode(stretchButtonPin,INPUT);
  digitalWrite(stretchButtonPin,HIGH);
}

void loop(){
  pot1val = analogRead(pot1pin);
  pot1val = map(pot1val, 0, 1023, 0, 179);
stretchButtonState = digitalRead(stretchButtonPin);
 if (stretchButtonState == LOW){
   for(servo1Position = pot1val; servo1Position < 179;
       servo1Position += 1)
  {
    myservo.write(servo1Position);
    delay(servoReflexDelay);
  }
 }
 else {
  myservo.write(pot1val);
  delay(servoDelay);
}
}
```

Working from top to bottom of the listing, we first load the servo library. Next, define the constants and pins used on the Arduino. In this example, we're using A0 as the analog input to read the position of potentiometer P1, digital pin D7 for switch S1, and digital pin D9 for `servo1pin`, the PWM output to drive the servo.

The `servoDelay` and `servoReflexDelay` variables define the responsiveness of the servo. Adjust these delay values to suit your servo. We found that the supplied values work well with intro-level HiTec and TowerPro servos. Increase the delay value a few milliseconds if your servo jitters at rest. For realism, use a `servoReflexDelay` time shorter than `servoDelay`; otherwise, the reflex will unfold in slow motion relative to the normal tracking speed of the servo.

In the setup area, first attach the servo to the servo object, as required by the servo library. Then establish the stretch-button pin as an input pin, with the internal pull-up enabled. Alternatively, you could use an external 10-kΩ, 1/8-W resistor tied to the 5-V supply, but it's easier in software, and there's really no benefit to the external hardware.

The main loop begins by reading the voltage from the wiper arm of potentiometer P1, which should be in the range of 0 to 5 V. Next, the `map` function scales the 0–1,023 value returned by input port A0 to fit the range of a typical analog hobby servo, 180 degrees. When button S1 is pressed, the normally high pin D7 is brought to ground potential, triggering the reflex.

Simulating the effect in real muscle, the magnitude of the reflex depends on the initial position of the servo. At one extreme servo position, akin to a muscle that is fully contracted, the reflex will be just noticeable. At the other extreme, akin to a muscle that is fully extended, the reflex will encompass most of the 180-degree range of the servo.

When you're debugging your system, it may be difficult to appreciate subtle changes in servo operation with seemingly major changes in coding. You can minimize debugging time by using the `Serial.println` function to track any or all variables in your program. For example,

```
Serial.println(FSRValue);
```

prints the value of the variable `FSRValue`. Just be sure to have your serial monitor window open and to add `Serial.begin` to your setup routine, as in

```
#define DEBUG
.
.
void setup(){
  #ifdef DEBUG
Serial.begin(9600);
#endif
}
```

The `#define DEBUG` statement should appear near the top of your program, where you can easily locate it and comment it out of the compile cycle when you're finished debugging. Also, if you opt to call the function with the default 9,600 baud, as in this example, then verify that the serial monitor is set to 9,600 baud.

Operation

This is a simulator or electromechanical approximation of what happens when you trigger a stretch reflex, as discussed earlier. In operation, you'll use potentiometer P1 to adjust the position of the servo arm and switch S1 to trigger the reflex. The reflex serves to rotate the servo arm away from or toward the striking object depending on where the switch is mounted. After a second or two, the servo returns to its original position, still under control of the potentiometer. To get things going, upload the software, and verify that you can control the servo throughout the full 180-degree range with potentiometer P1. Next, adjust the potentiometer so that the servo is midrange, and then tap S1. If all goes well, the servo will quickly rotate to one endpoint and back to the midpoint.

Next, you'll need to work the kinks out of the system. If your servo is jittery, then try different values for `servoDelay`. Smell something burning? If so, you probably melted the Arduino's onboard voltage regulator. It's also a good time to try different servos and types of potentiometers if you have them.

Modifications

This simple reflex arc simulator is fully functional and useful as in an android design. As with human reflexes, the servo operates normally until the reflex is triggered. There are, however, several simple modifications (mods) that you can make to provide more realistic and potentially more useful functionality. Follow along, in sequence, as we add to the hardware and code base of the simple reflex arc simulator.

Direction of Reflex

The first modification is to programmatically define the direction of the reflex. If you mounted switch S1 on the servo arm, then the reflex can either move the button toward or away from the striking object. If the direction of the reflex doesn't suit your needs, you could either move the switch or decrement the position variable down to zero. That is, in place of the increment code:

```
for(position = pot1val; position < 179; position += 1)
```

use this line to decrement the `position` variable:

```
for(position = pot1val; position >= 1; position -= 1)
```

Let's say that switch S1 is located in the toe area of a bipedal android. In general, an appropriate reflex would be to retract the foot away from an object that activates S1. Think of stubbing your toe and your normal reflex—before you start yelling niceties. However, there may be circumstances where the natural reflex is inappropriate, and you need to reverse the direction of the reflex under program control. Let's introduce a direction variable into the program, `reflexDirection`, and the subroutine `makeReflex`, as in Listing 1-2.

LISTING **1-2**   Arduino code for directing direction of reflex.

```
/*
Simple Reflex Arc Simulator with definable direction of reflex
Androids: Build Your Own Lifelike Robots by Bergeron/Talbot
Listing 1-2. See www.mhprofessional.com/Androids for fully
documented code
Arduino 1.0.1 environment
*/

#include <Servo.h>
Servo myservo;
const int pot1pin = A0;
const int stretchButtonPin = 7;
const int servo1pin = 9;

int pot1val;
int servoDelay = 15;
int servoReflexDelay = 2;
int stretchButtonState = 0;
int servo1Position = 0;
int reflexDirection = 0;

void setup(){
  myservo.attach(servo1pin);
  pinMode(stretchButtonPin,INPUT);
  digitalWrite(stretchButtonPin,HIGH);
}

void loop(){
  pot1val = analogRead(pot1pin);
  pot1val = map(pot1val, 0, 1023, 0, 179);
 stretchButtonState = digitalRead(stretchButtonPin);
 if (stretchButtonState == LOW){
   MakeReflex();
```

```
    }
  else {
   myservo.write(pot1val);
    delay(servoDelay);
 }
 }


/*
   ----------------------------------------------------
  makeReflex()
   ----------------------------------------------------
  */

void MakeReflex(){
   if (reflexDirection == LOW){
    for(servo1Position = pot1val; servo1Position < 179;
        servo1Position += 1){
     myservo.write(servo1Position);
     delay(servoReflexDelay);
     }
   }
   else{
    for(servo1Position = pot1val; servo1Position>=1;
        servo1Position -= 1){
     myservo.write(servo1Position);
     delay(servoReflexDelay);
     }
   }
  }
```

The subroutine `makeReflex` replaces the code to increment the servo within the main loop. By changing the value of variable `reflexDirection` from 0 to 1, either manually or programmatically, you can reverse the direction of the reflex. Don't forget, the documented listing is available online.

Absolute Refractory Period

At this point, our simple reflex arc simulator has a natural absolute refractory period, the period of time when it is nonresponsive to switch S1 as a result of the electrical and mechanical design of the servo and mechanical attachments. Because no two servos are identical, it's a good idea to define an absolute refractory period that we can control programmatically to compensate for differences in responsiveness between servos. This sort of tuning becomes important when multiple limbs and even more servos are involved.

Let's define a constant `absoluteRefractory` to represent the absolute refractory period, the time, in milliseconds, that the reflex cannot be triggered by S1. Otherwise, the servo operates normally, with position under control of potentiometer P1. We could call the `delay()` function after a reflex has executed to provide a refractory period, but a better approach is shown in Listing 1-3.

LISTING 1-3   Arduino code for simple reflex arc with absolute refractory period.

```
/*
Simple Reflex Arc Simulator with absolute refractory period
Androids: Build Your Own Lifelike Robots by Bergeron/Talbot
Listing 1-3. See www.mhprofessional.com/Androids for fully
documented code
Arduino 1.0.1 environment
*/

#include <Servo.h>
Servo myservo;
const int pot1pin = A0;
const int stretchButtonPin  = 7;
const int servo1pin = 9;

int pot1val;
int servoDelay  = 5;
int servoReflexDelay  = 2;
int stretchButtonState  = 0;
int servo1Position = 0;
int reflexDirection = 0;
long previousMillis = 0;
long absoluteRefractory = 5000;

void setup() {
  myservo.attach(servo1pin);
  pinMode(stretchButtonPin,INPUT);
  digitalWrite(stretchButtonPin,HIGH);
}

void loop() {
  pot1val = analogRead(pot1pin);
  pot1val = map(pot1val, 0, 1023, 0, 179);
  unsigned long currentMillis = millis();
 stretchButtonState = digitalRead(stretchButtonPin);
 if ((stretchButtonState == LOW)&&(currentMillis - previousMillis >
     absoluteRefractory)){
```

```
    makeReflex();
    previousMillis = currentMillis;
  }
  else {
   myservo.write(pot1val);
   delay(servoDelay);
  }
}
}
/*
  ------------------------------------------------------
  makeReflex()
  ------------------------------------------------------
  */

void MakeReflex(){
  if (reflexDirection == LOW){
    for(servo1Position = pot1val; servo1Position < 179;
servo1Position += 1)
      {
     myservo.write(servo1Position);
     delay(servoReflexDelay);
     }
  }
  else{
   for(servo1Position = pot1val; servo1Position>=1;
      servo1Position -= 1){
    myservo.write(servo1Position);
    delay(servoReflexDelay);
    }
  }
}
```

Because we want to disable the reflex until the absolute refractory time has passed, we need to measure time. This is accomplished by using the `millis()` function, which returns the number of milliseconds that have passed since power was last applied to the Arduino. The variables `previousMillis` and `currentMillis` are used to store the results of the `millis()` function at different points in the program, thereby providing a measure of the time interval.

Although this approach is more complicated than simply using the `delay()` function, the setting, checking, and resetting of elapsed time in milliseconds have the advantage that they do not stop the processor. The `delay()` function essentially suspends the operation of the Arduino, making it blind to sensor activity.

Relative Refractory Period

Adding a relative refractory period to the simple reflex arc simulator, in which the stimulus required to elicit a reflex diminishes with time since the last reflex, not only adds realism to the reflex activity, but it's also a good excuse to add some additional hardware to the mix. Specifically, we need to replace switch S1 with an analog sensor that can respond in varying degrees to a stimulus.

Going for a minimalist approach, we selected a small force-sensing resistor, FSR1, available from Sparkfun. The resistance varies from about 1 MΩ at rest to 1 kΩ with the tap of a finger. This makes a good detector of stimulus level because it's fairly sensitive (100g), and the price is right. There is a peel-and-stick rubber backing that makes mounting a snap. The only real limitation of this sensor is the leads. Don't even think about soldering them. Instead, use a wire wrap tool or, if you have one, a Grove terminal block to connect the sensor to an Arduino. Figure 1-7 shows a close-up of a force-sensing resistor.

There's nothing special about this particular force-sensing resistor—feel free to substitute whatever you have on hand—from a linear potentiometer with spring return to a pair of stripped copper wires embedded in a 1-inch cube of conductive foam. The caveat is that you should use a sensor that's fairly sensitive. A sensor that requires you to hit it with a hammer isn't going to work well with our setup.

An alternative to a resistive sensor is a piezo pickup. While sensitive to finger taps, piezo pickups have a nonlinear output, so discriminating between strike intensities would be more difficult. You could also use a strain gauge for more

FIGURE 1-7 Force-sensing resistor.

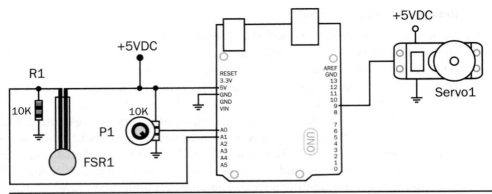

Figure 1-8 Circuit to explore relative refractory period. P1 controls servo position, and FSR1 initiates the reflex.

accurate pressure readings, but the added accuracy comes at a considerable cost. For now, let's stick with the force-sensitive resistor.

Follow along on Figure 1-8 for the hardware additions. First, let's create a voltage divider with a 10-kΩ resistor R1 and the force-sensing resistor FSR1. One side of the sensor is connected to 5 Vdc, and the other side is connected to both A1 and the 10-kΩ resistor. The other side of R1 is at ground potential.

With no pressure on the force-sensing resistor, essentially no voltage appears across the 10-kΩ resistor. Tap the sensor with your finger, and the sensor resistance should drop from over 1 MΩ to perhaps 2-kΩ, causing a greater proportion of the 5 V to drop across the resistor. Recall the voltage-divider equation, where the voltage drop across one resistor of a series pair is proportional to the value of that resistor divided by the sum of both resistors. In this case, voltage across R1 (VR1) is equal to the supply voltage (5 V) times the value of R1 divided by the sum of R1 and FSR1:

$$V_{R1} = 5 \text{ V} \times R1/(R1 + FSR1)$$

So, with no load on FSR1, we have

$$V_{R1} = 5 \text{ V} \times 10,000/(10,000 + 1,000,000) = 0.05 \text{ V}$$

And at full load on FSR1, we have

$$V_{R1} = 5 \text{ V} \times 10,000/(10,000 + 1,000) = 4.55 \text{ V}$$

The code enabling a relative refractory period is shown in Listing 1-4. Note the addition of a new reset subroutine.

LISTING 1-4   Arduino code for simple reflex arc simulator with absolute and relative refractory periods.

```
/*
Simple Reflex Arc Simulator with absolute and relative refractory
periods
Androids: Build Your Own Lifelike Robots by Bergeron/Talbot
Listing 1-4. See www.mhprofessional.com/Androids for fully
documented code
Arduino 1.0.1 environment
*/

#include <Servo.h>
Servo myservo;
const int pot1pin = A0;
const int FSRPin = A1;
const int servo1pin = 9;

int pot1val;
int servoDelay = 5;
int servoReflexDelay = 2;
int stretchButtonState = 0;
int servo1Position = 0;
int reflexDirection = 0;
long absoluteMillis = 0;
long absoluteRefractory = 1000;
unsigned long currentMillis = millis();
unsigned long relativeMillis = millis();
int FSRValue = 0;
int FSRMaxVal = 0;
int reflexFlag = 0;
int noiseLevel = 175;
int triggerVal = 0;
int sensorRange =1000;

void setup() {
  myservo.attach(servo1pin);
}

void loop(){
  pot1val = analogRead(pot1pin);
  pot1val = map(pot1val, 0, 1023, 0, 179);
  currentMillis = millis();
```

```
       FSRValue = 0;

   if (currentMillis - absoluteMillis > absoluteRefractory){
     FSRValue = analogRead(FSRPin);
   }

   if (FSRValue>noiseLevel) {
     FSRMaxVal = max(FSRMaxVal,FSRValue);
     reflexFlag =1;
   }

   if ((FSRValue<noiseLevel) && (reflexFlag == 1)){
   triggerVal = max (noiseLevel, (sensorRange-((currentMillis -
                    relativeMillis )/4)));
     if (FSRMaxVal > triggerVal) {
         makeReflex();
         reset();
       }
       else{
         reset();
       }
   }
    else {
       myservo.write(servo1Position);
       delay(servoReflexDelay);
   }
   }

   /*
    ----------------------------------------------------------
    makeReflex()
    reset()
    ----------------------------------------------------------
    */

   void MakeReflex(){
     if (reflexDirection == LOW){
       for(servo1Position = pot1val; servo1Position < 179;
          servo1Position += 1){
        myservo.write(servo1Position);
        delay(servoReflexDelay);
       }
```

```
    }
    else{
     for(servo1Position = pot1val; servo1Position>=1;
         servo1Position -= 1) {
      myservo.write(servo1Position);
      delay(servoReflexDelay);
      }
    }
  }

void reset() {
    absoluteMillis = currentMillis;
    relativeMillis = currentMillis;
    reflexFlag =0;
    FSRMaxVal =0;
}
```

As illustrated in the following code snippet, the program won't read the force sensor, much less calculate the refractory period, if the system is still in the absolute refractory period.

```
if (currentMillis - absoluteMillis > absoluteRefractory){
FSRValue = analogRead(FSRPin);
}
```

Once the system is past the absolute refractory period, code relevant to the relative refractive period kicks in. From the listing, you can see that the variable `sensorRange` defines the range of the force sensor and therefore the relative refractory period.

As illustrated in Figure 1-9, the exponentially decreasing trigger level for the relative refractory period is approximated by the standard equation for a straight line, $y = mx + b$, where x and y are axes, m is the slope of the line, and b is a constant. For this program, we use a slope of ¼, or 0.25, and a value of 1,000 for b, which is represented by the value of `sensorRange`.

The relevant code involves first determining the value of the minimum trigger level, `triggerVar`, as a function of time and the system noise level, `noiseLevel`. Although there isn't noise in the usual sense, the `noiseLevel` variable is a convenient means of approximating the shift in the minimum trigger conditions for a reflex.

Furthermore, because periodically sampling the force-sensing resistor returns a string of values that peak somewhere in midstream, we use a peak-detector algorithm, using the `max` function. Once the minimum trigger level is established,

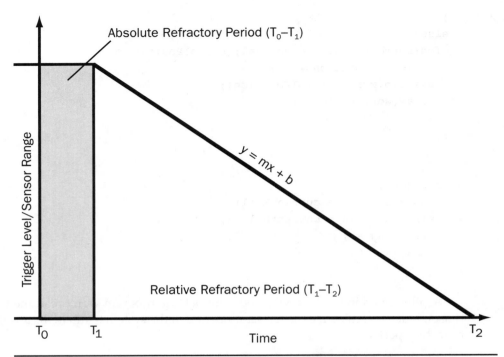

FIGURE 1-9 Relationship between absolute and relative refractory periods. The straight-line decrease in trigger level approximates the exponential decrease observed in biological systems.

it's a simple matter of comparing it with the peak value of the signal returned by the force-sensing resistor and stored in the variable `FSRMaxVal`.

To experience the relative refractory period, with one hand, operate the servo by rotating potentiometer P1, and with the other hand, tap the force-sensing resistor to fire the reflex. You'll have to hit FSR1 hard to elicit a reflex in the first few milliseconds after a reflex and less so as the milliseconds and seconds pass by. If you plan to verify this, it's a good idea to mount the force-sensing resistor FSR1 on your desktop instead of on the servo horn.

Brain Freeze

Recall that the brain has a negative effect on reflex arcs. This can be a great thing for your android when the context suggests that reflexes are a bad thing—think a battle android hiding in the jungle with some animal gnawing on its ankle—better to be perfectly still than to zap the pest and be spotted by the enemy. Let's simulate cortical inhibition by reintroducing S1 using the same pin connection as before, as in Figure 1-10.

Building on the work of absolute and relative inhibition, the related code to support inhibition of the reflex arc is shown in Listing 1-5. We're introducing a new

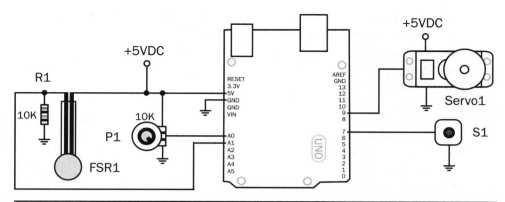

FIGURE 1-10   Setup for inhibition experiment. P1 controls servo position, FSR1 initiates the reflex, and S1 inhibits the reflex.

variable associated with activation of S1, `globalInhibition`, and using it to increase the value of the noise-level variable `noiseLevel`, which is the minimum signal level required to activate a reflex arc.

LISTING 1-5   Arduino code for reflex inhibition of a reflex arc.

```
/*
Simple Reflex Arc Simulator with refractory periods and inhibition
Androids: Build Your Own Lifelike Robots by Bergeron/Talbot
Listing 1-5. See www.mhprofessional.com/Androids for fully
documented code
Arduino 1.0.1 environment
*/

#include <Servo.h>
Servo myservo;
const int pot1pin = A0;
const int FSRPin = A1;
const int servo1pin = 9;
const int inhibitionPin = 7;

int pot1val;
int servoDelay = 5;
int servoReflexDelay = 2;
int inhibitionState = 0;
int servo1Position = 0;
int reflexDirection = 0;
```

```
long absoluteMillis = 0;
long absoluteRefractory = 1000;
unsigned long currentMillis = millis();
unsigned long relativeMillis = millis();
int FSRValue = 0;
int FSRMaxVal = 0;
int reflexFlag = 0;
int noiseLevel = 175;
int triggerVal = 0;
int sensorRange =1000;
int globalInhibition = 250;

void setup() {
  myservo.attach(servo1pin);
}

void loop(){
  pot1val = analogRead(pot1pin);
  pot1val = map(pot1val, 0, 1023, 0, 179);
  currentMillis = millis();
  FSRValue = 0;
  inhibitionState = digitalRead(inhibitionPin);

if (currentMillis - absoluteMillis > absoluteRefractory) {
  FSRValue = analogRead(FSRpin);
}

if (FSRValue> (noiseLevel + (inhibitionState * globalInhibition))){
  FSRMaxVal = max(FSRMaxVal,FSRValue);
  reflexFlag =1;
}

if ((FSRValue<noiseLevel) && (reflexFlag == 1)){
triggerVal = max (noiseLevel, (sensorRange-((currentMillis -
                   relativeMillis )/4)));
  if (FSRMaxVal > triggerVal) {
      makeReflex();
      reset();
    }
    else{
      reset();
    }
```

```
  }
  else {
     myservo.write(pot1val);
     delay(servoDelay);
  }
}
/*
 ------------------------------------------------------------
 makeReflex()
 reset()
 ------------------------------------------------------------
 */

void MakeReflex(){
   if (reflexDirection == LOW){
     for(servo1Position = pot1val; servo1Position < 179;
servo1Position += 1){
      myservo.write(servo1Position);
      delay(servoReflexDelay);
      }
   }
   else{
     for(servo1Position = pot1val; servo1Position>=1;
        servo1Position -= 1){
      myservo.write(servo1Position);
      delay(servoReflexDelay);
      }
   }
 }

void reset() {
    absoluteMillis = currentMillis;
    relativeMillis = currentMillis;
    reflexFlag =0;
    FSRMaxVal =0;
}
```

When S1 is depressed, connecting inhibitionPin D7 to ground, noiseLevel is raised by 250, significantly raising the minimum pressure-sensor value required to trigger the makeReflex() function.

If you'd like to experiment with a more realistic simulation of cortical inhibition, substitute a potentiometer for the momentary contact switch S1. You'll quickly notice that tuning the system becomes more difficult because of the interplay of relative refractory period and small to modest amounts of inhibition.

As in humans, inhibition of specific reflexes can come in handy at times. For example, if your android is walking across a yard or field and the "toe" sensors are continually activated by tall grass, the associated reflexes should be inhibited until the terrain improves.

Adrenaline Rush

In a life or death situation, an android may have to forgo normal operating constraints and go for it. Let's say that there's 30 seconds for your android to get to the quadcopter rescue drone before it takes off, and your android has to pull out the stops and sprint for it. We'll simulate the adrenaline rush called for in this situation by adding a second switch, S2, to digital pin D6 of the Arduino, as in Figure 1-11.

Given that the overall effect of an adrenaline rush is practically the mirror opposite of cortical inhibition, we'll use a parallel code structure with a negative instead of positive effect on the effective noise level, as shown in Listing 1-6.

LISTING **1-6**   Arduino code to add a simulated adrenaline rush to the reflex arc.

```
/*
Simple Reflex Arc Simulator with refractory periods, inhibition,
and excitation
Androids: Build Your Own Lifelike Robots by Bergeron/Talbot
Listing 1-6. See www.mhprofessional.com/Androids for fully
documented code
Arduino 1.0.1 environment
*/
```

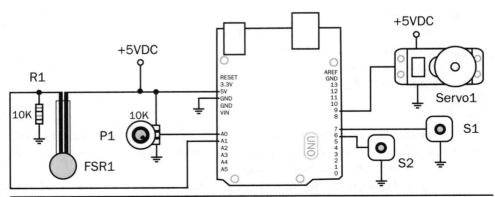

FIGURE **1-11**   Addition of switch S2 to trigger a simulated adrenaline rush. P1 controls servo position, FSR1 initiates the reflex, and S1 inhibits the reflex.

```
#include <Servo.h>
Servo myservo;
const int pot1pin = A0;
const int FSRPin = A1;
const int servo1pin = 9;
const int excitationPin = 6;
const int inhibitionPin = 7;

int pot1val;
int servoDelay = 5;
int servoReflexDelay = 2;
int inhibitionState = 0;
int excitationState = 0;
int servo1Position = 0;
int reflexDirection = 0;
long absoluteMillis = 0;
long absoluteRefractory = 1000;
unsigned long currentMillis = millis();
unsigned long relativeMillis = millis();
int FSRValue = 0;
int FSRMaxVal = 0;
int reflexFlag = 0;
int noiseLevel = 175;
int triggerVal = 0;
int sensorRange =1000;
int globalInhibition = 250;
int globalExcitation = -175;

void setup() {
  myservo.attach(servo1pin);
}

void loop(){
  pot1val = analogRead(pot1pin);
  pot1val = map(pot1val, 0, 1023, 0, 179);
  currentMillis = millis();
  FSRValue = 0;
  inhibitionState = digitalRead(inhibitionPin);
if (currentMillis - absoluteMillis > absoluteRefractory){
  FSRValue = analogRead(FSRPin);
}
```

```
if (FSRValue> (noiseLevel + (inhibitionState * globalInhibition) +
    (excitationState * globalExcitation))) {
  FSRMaxVal = max(FSRMaxVal, FSRValue);
  reflexFlag =1;
}

if ((FSRValue<noiseLevel) && (reflexFlag == 1)){
triggerVal = max (noiseLevel, (sensorRange-((currentMillis -
                 relativeMillis )/4)));
  if (FSRMaxVal > triggerVal){
      makeReflex();
      reset();
    }
    else {
      reset();
    }
}
 else {
    myservo.write(pot1val);
    delay(servoDelay);
}
}

/*
 --------------------------------------------------------
 makeReflex()
 reset()
 --------------------------------------------------------
 */
void makeReflex(){
  if (reflexDirection == LOW){
   for(servo1Position = pot1val; servo1Position < 179;
      servo1Position += 1) {
    myservo.write(servo1Position);
    delay(servoReflexDelay);
    }
  }
  else{
   for(servo1Position = pot1val; servo1Position>=1;
      servo1Position -= 1) {
    myservo.write(servo1Position);
    delay(servoReflexDelay);
```

```
      }
    }
  }

void reset() {
    absoluteMillis = currentMillis;
    relativeMillis = currentMillis;
    reflexFlag =0;
    FSRMaxVal =0;
}
```

Essentially, `excitationState` and `globalExcitation` are used just as the variables `inhibitionState` and `globalInhibition` to simulate inhibition. The only difference here is that `globalExcitation` contributes a negative value to `noiseLevel`, thereby lowering the threshold necessary to trigger a reflex arc. Although you can raise the noise level to anything you desire, you can decrease the noise level only so far. As you approach the real noise level—somewhere above zero—your android will be "on edge." The slightest vibration is likely to set off a cascade of reflex arcs—not necessarily the best personality for an android.

Note that the method of calculating the contribution of inhibitory and excitatory factors to the trigger level relies on a simple algebraic summation of factors:

```
(noiseLevel + (inhibitionState * globalInhibition) +
(excitationState * globalExcitation))
```

You can build on this elementary form of sensor fusion by assigning different weights to inhibition and excitation values. Using this approach, you could make inhibition three times as significant as excitation, or vice versa, with the addition of a constant multiplier. Or you can track previous values as part of a predictive algorithm. In short, don't be lulled into the easy combination of sensor values here—experiment with methods of fusing and combining sensor readings to achieve the behaviors you're after.

As with the preceding simulation of inhibition, feel free to substitute a potentiometer for switch S2. Doing so can give you a feel for the interaction of competing sensor inputs combined with the excitatory and inhibitory effects. Real-time manipulation of inhibition and excitation of reflex arcs with potentiometers can save time when you're tuning a complex system of multiple servos and sensors.

One of your android's arms may have a relatively sluggish servo, for example, and values for refractory period that work with servos on the other arm won't work. Instead of tweaking variables in an editor and then recompiling, you can adjust a potentiometer in real time and observe the results. Once you determine the optimal values, you can easily remove the components and code in their equivalent values.

Of course, you can programmatically change the relative levels of inhibition and stimulation triggered by S1 and S2, and as noted earlier, the overhead of switches is less than reading a pair of potentiometers. Switching is useful in situations where you have known differences in operating parameters. For example, you might have two battery packs, one standard and one extended (and heavy). Switching is a great way to modify the reflexes to presets to compensate for the extra weight and change in center of gravity.

Additional Servos

Now that you've explored the basics, we can hang more effectors (servos) on the system. Let's add a second servo, as in Figure 1-12, to simulate either a second muscle or a second bundle of muscles fibers within the same muscle.

The code to support an additional servo is shown in Listing 1-7. In this example, the second servo mirrors the movement of the first. Otherwise, we'd have to add a second potentiometer and all the associated code.

Listing **1-7**   Arduino code to support a second servo.

```
/*
Simple Reflex Arc Simulator with refractory periods/inhibition/
excitation and additional servo
Androids: Build Your Own Lifelike Robots by Bergeron and Talbot
Listing 1-7. See www.mhprofessional.com/Androids for fully
documented code
Arduino 1.0.1 environment
*/
```

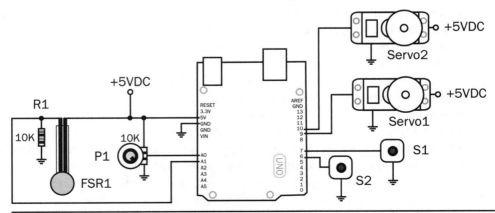

Figure **1-12**   Addition of a second servo, Servo2, to simulate a second muscle or second bundle of muscle fibers in the same muscle. P1 controls servo position, FSR1 initiates the reflex, S1 inhibits the reflex, and S2 facilitates the reflex.

```
#include <Servo.h>
Servo myservo;
Servo myservo2;
const int pot1pin = A0;
const int FSRPin = A1;
const int servo1pin = 9;
const int servo2pin = 10;
const int excitationPin = 6;
const int inhibitionPin = 7;

int pot1val;
int servoDelay = 5;
int servoReflexDelay = 2;
int inhibitionState = 0;
int excitationState = 0;
int servo1Position = 0;
int reflexDirection = 0;
long absoluteMillis = 0;
long absoluteRefractory = 1000;
unsigned long currentMillis = millis();
unsigned long relativeMillis = millis();
int FSRValue = 0;
int FSRMaxVal = 0;
int reflexFlag = 0;
int noiseLevel = 175;
int triggerVal = 0;
int sensorRange = 1000;
int globalInhibition = 250;
int globalExcitation = -175;

void setup() {
  myservo.attach(servo1pin);
  myservo2.attach(servo2pin);
}

void loop(){
 pot1val = analogRead(pot1pin);
  pot1val = map(pot1val, 0, 1023, 0, 179);
  currentMillis = millis();
  FSRValue = 0;
  inhibitionState = digitalRead(inhibitionPin);
if (currentMillis - absoluteMillis > absoluteRefractory){
```

```
    FSRValue = analogRead(FSRPin);
  }
  if (FSRValue> (noiseLevel + (inhibitionState * globalInhibition) +
      (excitationState * globalExcitation))){
  FSRMaxVal = max(FSRMaxVal,FSRValue);
    reflexFlag = 1;
  }
  if ((FSRValue<noiseLevel) && (reflexFlag == 1)) {
  triggerVal = max (noiseLevel, (sensorRange-((currentMillis -
                    relativeMillis )/4)));
    if (FSRMaxVal > triggerVal) {
        makeReflex();
        reset();
      }
      else {
        reset();
      }
  }
   else {
      myservo.write(pot1val);
      delay(servoDelay);
  }
  }


  /*
   ------------------------------------------------------------
   makeReflex()
   reset()
   ------------------------------------------------------------
   */

  void makeReflex(){
    if (reflexDirection == LOW){
     for(servo1Position = pot1val; servo1Position < 179;
         servo1Position += 1) {
      myservo.write(servo1Position);
      delay(servoReflexDelay);
      }
    }
    else{
     for(servo1Position = pot1val; servo1Position>=1;
         servo1Position -= 1) {
```

```
    myservo.write(servo1Position);
    delay(servoReflexDelay);
    }
  }
}

void reset() {
    absoluteMillis = currentMillis;
    relativeMillis = currentMillis;
    reflexFlag = 0;
    FSRMaxVal = 0;
}
```

The critical points are first creating a second servo object by declaring `myservo2` and then associating the second servo with a pin number in the `setup()` routine. Moving the two servos in synchronization is accomplished by calling the `write` function from within the `makeReflex()` subroutine.

You can modify the response of the second servo by adding additional controls and code to specify refractory period and inhibitory period and simulate an adrenaline rush. Consider, as an exercise, inhibiting one servo when the other is active—simulating the biceps/triceps or quad/hamstring reflexes. The challenge that you'll shortly encounter, however, is the limited number of analog ports available on the Arduino for sensors without going to measures such as port multiplexing.

Additional Computational Hardware

If your little Arduino Uno is overtaxed, you can turn to a dedicated servo controller to free up ports and processing for sensors and other tasks. Moreover, a dedicated servo controller enables you to configure the PWM pulse rate and pulse width, assign speed and acceleration for each servo, and assign startup position, all through a standard serial connection. You can also program sequences, such as reflexes, and trigger them with a single serial command from the Arduino.

The Maestro servo controllers from Pololu have a rich feature set, are available in a variety of form factors, and can be used with a free configuration control program. Parallax sells a propeller-based controller that is fully programmable and loaded with hardware and connectors for 16 servos. The downside of these or other dedicated servo controllers is cost—expect to spend about as much as for an Arduino.

Parallax also sells an in-between option that's not quite a controller but that nonetheless can free up Arduino resources. Parallax's ServoPAL is an inexpensive, tiny module that plugs between servo plugs and wires. It simply repeats the first set of PWM pulses it receives from the Arduino—all without programming.

Additional Sensors

Any sensor that provides a digital output can be substituted for the momentary contact switches in the example circuits, often with minimal recoding. For example, a mercury tilt switch, magnetic reed switch, and phototransistor can be connected to a digital input port with no recoding. Other sensors with digital output, such as infrared (IR) and ultrasonic range finders, will require modest coding.

The best part about using digital sensors is that the Arduino, like most microcontrollers, has lots of digital ports. However, when things get complicated— meaning more than a pair of servos and a couple sensors—it's a good idea to isolate the power and signal lines of the sensors, servos, and microcontroller. Use separate supplies for the servos and sensors, and isolate the signal lines by coupling the sensors to the microcontroller through an optical isolator, such as the 4N35.

Analog sensors generally offer more flexibility than their digital counterparts. However, they usually impose more overhead for processing than digital sensors and, of course, demand one of the limited analog inputs. Heat sensors, pressure sensors, Hall effect sensors, and sound sensors are all worth investigating in the context of reflex arcs.

Gremlins

As robotic circuits go, you can't get much simpler than the examples discussed here. You probably noticed that the examples don't use interrupts or fancy algorithms. It's no doubt possible to reduce much of the code to a few lines that perform elegant bit-flipping operations, and you're welcome to do so. For our purposes, though, simple integer operations get the job done.

However, even simple robotics is by definition multidisciplinary and problematic in practice. As such, expect glitches from time to time. Let's say that a servo is acting erratically, sometimes providing a smooth reflex and other times just jittering for a few seconds. The code may be perfect and the wiring impeccable. These are necessary but insufficient conditions for a fully functioning robot.

Zeros and ones are fairly standard in programming, but every physical device is unique, even down to a simple momentary contact switch. Furthermore, the operation of even the simplest device approaches chaos during the first few milliseconds of operation. Take our momentary button S1. When the button is pressed, the electric contacts slam against each other and bounce around, arcing and making a physical racket a hundred times or so in the first few milliseconds. If the microcontroller samples the electrical connection frequently enough, it will record conflicting data—open one read and closed the next two or three reads.

You can address the fluctuating readings by debouncing the switch, either in hardware or in software. The hardware approach, adding a resistor and capacitor in the form of a resistor-capacitor (RC) filter, saves processor time but requires circuit board space, and there's the added expense for components. The software approach, simply waiting for the switch to settle down, can be accomplished with a single *delay* statement. However, waiting essentially shuts down the microcontroller—not the best approach when multiple sensors are involved in a reflex arc.

Another area to check when things just don't seem to work properly is the power supply. Are glitches from the servo making their way into the power supplying the microcontroller? If you're not sure, consider installing bypass capacitors at the point power enters the Arduino and using a separate supply for servos and the microprocessor.

The bottom line is that even simple components aren't perfect, much less complex sensors. When you're debugging these and other robotic circuits and software, do your research and understand how components operate, especially at turn-on and turn-off, when algorithms tend to fail.

Search Terms

If you'd like to dive deeper into the application of reflexes in robotics, try the following search terms for your browser:

- Braitenberg vehicle
- Subsumption architecture
- Sensor fusion

CHAPTER 2

Behavior Chains

For most people, most of the time, creativity and human intelligence are overrated and underutilized. Humans are creatures of habit—that is, behavior chains that are played and replayed day after day. True, that guy driving the 18-wheeler behind you might be composing a sonata or agonizing over the significance of the Higgs boson, but from an observable, behavioral perspective, he might as well be deliberating over what to have for lunch. All that matters to his employer is that he delivers the goods intact and on time. And all you should care about is that he knows the difference between the brake and gas pedals. All else is moot.

It's the same with androids. It doesn't matter what they're thinking as long as they get the job done. If you're designing an android bodyguard, what's the point of enabling it to dream about, say, electric sheep? If anything, this sort of mental activity likely detracts from the work to be done. And most of the useful work performed by humans—and, eventually, androids—tends to be predictable, repetitive patterns of behavior. The linked patterns or sequences of stimulus-driven, hormonally and cortically mediated reflexes that constitute these behaviors are called *behavior chains*.

This chapter explores simple behavior chains and how they can be modeled and harnessed to perform seemingly complex tasks with modest programming and computational overhead. Using an Arduino, a robot arm, and a few sensors, we'll illustrate how to enable an android with autonomous, predictable behaviors that leverage the concept of the reflex arc introduced in Chapter 1.

Biological Basis

Humans engage in subconscious and locally mediated activity most of the time, and these activities are built on sequences or chains of reflex arcs. Figure 2-1 shows one of many possible chain configurations. From Wait State 1, Trigger Event A initiates Reflex A, and depending on whether Condition 1 or 2 is present, the behavior chain links to either Reflex B or Reflex D. Reflex B is under the inhibitory influence of the brain and the excitatory influence of adrenaline. Reflex B links to Reflex C, which links back to the original Wait State 1.

Again, referring to Figure 2-1, if Reflex A occurs under Condition B, then Reflex A is followed by Reflex D, which links to Wait State 2. This chain stays in the wait state until Trigger Event B, which links to Reflex C, which links back to the original Wait State 1. However, depending on the refractory period of Reflex A, it's possible that another Trigger Event A will set the behavior chain in motion before a Trigger Event B occurs. This hypothetical—and admittedly dreadful—discussion of A's and B's should have more meaning when we discuss the behavior chain in the context of specific reflexes. For now, just get comfortable with Figure 2-1.

As in the military, behavior chains are primarily hurry up and wait affairs, in that nothing happens for relatively great expanses of time until the trigger event(s). The evolutionary goal, of course, is to provide leisure time for that big cerebral cortex so that it can devise strategies to dominate *Call of Duty*. Whereas few of us have identical cortical workloads, many behavior chains are universal and often have an equivalent in other life forms. Consider the following account of the history of one such artificial life form, the microprocessor.

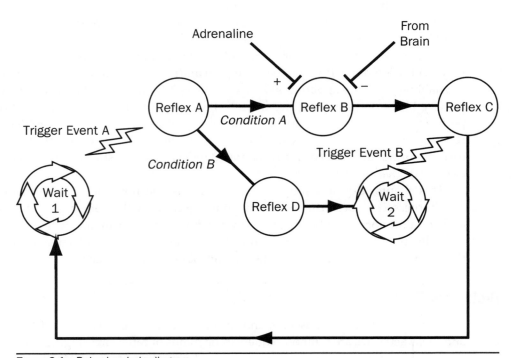

Figure 2-1   Behavior-chain diagram.

Mouse Maze

When Intel introduced the microprocessor in 1971, it faced an uphill battle to get the pocket-protector crowd to change its analog approach to problem solving. To increase awareness of its new digital device, Intel held a contest in which a $10,000 first prize awaited the team that created an electronic mouse—using the Intel chip—that made it out of an Intel-specified maze in the shortest time.

Physically, this was more like a rat contest because the electronic mice involved were about the size of a shoe. There were some practical constraints, such as the size of the electric motors used to drive the wheels, the size of the battery, and the amount of random access memory (RAM) on board. However, for the most part, the inner workings of the mouse were up to the team's imagination and ingenuity. As with contemporary robots, the real challenge was devising an elegant solution that could fit within the computational and memory constraints of the microcontroller.

Contestants used a variety of approaches, from employing infrared (IR) light sensors and switches on all sides of the mouse body to whiskers that detected contact with the walls of the maze. Most of the designs attempted to create a map of the maze in the mouse's onboard RAM. However, the winning team took a simpler, more elegant approach. The team took the time to study the habits—that is, the behavior chains—of real mice and how the rodents managed to navigate the crumbling walls of buildings in every major city on the planet.

The winning team learned that mice come in two basic varieties, either left-handed or right-handed. Left-handed mice, when faced with a choice of turning left or right at the end of a tunnel, favor the left. Right-handed mice, in contrast, favor taking the right when presented with the same choice. It turns out that the strategy of consistently taking the right or left turn when presented with a left-right choice always brings a mouse out of a maze regardless of the size or complexity of the maze—assuming that the mouse has adequate food and water for the journey.

As you've probably guessed by now, the winning team created a right-handed mouse. The team didn't need the new digital microprocessor but incorporated it into the design in order to satisfy the contest rules. Team members used the microprocessor to monitor a switch attached to a plastic whisker on the right-hand side of the mouse. When the whisker lost contact with the right wall, the robot turned right until it regained contact with the wall. In this way, it hugged the right-hand wall and eventually made its way out of the maze. The other robots, some using complex machine-learning algorithms, were either trapped in the maze or eventually stumbled out.

Clearly, the thinking machines were outdone by the doing machine. Modeling a simple behavior chain—acting consistently rather than agonizing—beat all that digital processing power. Want an affordable android that can escape one of those topiary mazes without access to a Watson supercomputer or an electric hedge trimmer? Then forget artificial intelligence (AI)—go for *intelligent action* (IA).

Intelligent Action and Behavior Chains

A behavior chain is a sequence of reflexes that are fired off in response to some stimulus. Slapping your arm in response to a mosquito bite, scratching a patch of skin in response to an itch, recoiling in response to a too-hot stovetop, looking toward a flash of light, and cringing in response to a loud sound are simple behavior chains based on a few reflexes. Many common chains involve numerous, intricately interwoven reflexes, often with potentially significant input from the brain and endocrine system.

It turns out that our big brains are good for something other than pondering the nature of π—pattern recognition. And our brains are especially good at recognizing threatening and potentially lethal patterns. We're hardwired to instantly recognize the one frowning face—the one potential threat—in a sea of smiling faces. Similarly, someone might ignore one bee but take off, adrenaline coursing through their veins, when confronted with a swarm.

Humans are born with innate behavior chains or instincts, but many are not apparent until the nervous system matures. Other behavior chains are formed only through experience. For example, most infants instinctively chug down ice cream but spit up creamed spinach. And, probably indicative of our ancestors brachiating in the forest canopy, humans don't develop fear of heights until they are several years old. Similarly, most toddlers are visibly startled by a loud noise, but toddlers who fail to develop the specific Pavlovian behavior chain of getting off the road and onto the sidewalk don't contribute to the gene pool.

In addition to our shared behavior chains, we have personal behavior chains that define us as individuals. The main personality axes, equally applicable to humans, androids, and toasters, are dominance, conscientiousness, (emotional) stability, and openness (Figure 2-2). And it's the relative expression of each trait that defines a particular personality at a point in time. For example, the personality of a stereotypical servant or service robot would be submissive, conscientious, emotionally stable, and open. On the other hand, a stereotypical soldier would score high in dominance, low in conscientiousness and openness, and only modestly positive in stability.

Personality profiles evolve over time, with changes in the environment and, when a human is involved, with age and health status. The personality of an older human is typically less sanguine—that is, less open and less dominant. Disease, including that of the mind, can lead to emotional instability and lack of conscientiousness. We'll delve into the expected effects of aging on androids in more detail in Chapter 8.

Just Noticeable Difference

A consideration related to the triggering of behavior chains is a property of human perception called *just noticeable difference* (JND). And it's exactly what the name suggests—it's the change in volume, pitch, temperature, light intensity, or other

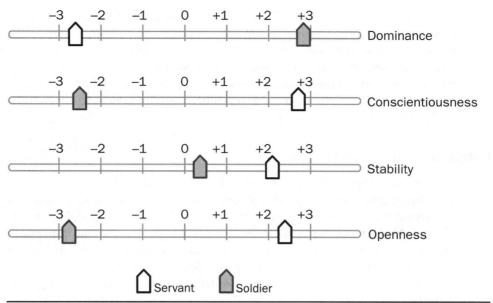

FIGURE 2-2   Major personality axes, with personality profiles of a stereotypical servant and soldier.

stimuli that's noticeable at least 50 percent of the time (hence the term *just*). For most people and most stimuli, the ratio of the JND level of, say, volume to the original intensity is constant—the *Weber constant*.

As a practical example, if you can detect a difference in volume about half the time when someone moves the volume control on your stereo amplifier from 2 to 4, then you should be able to detect a difference in volume about half the time when someone cranks it up from 9 to 11. This assumes an audio taper potentiometer and that the numerical labels are spaced equally around the pot knob.

Relevance to Android Designs

Just as behavior chains ultimately define the personality of an individual, behavior chains define the personality of an android—or even a machine as simple as a robot arm. This shouldn't be surprising because we anthropomorphize the world around us by nature. We routinely ascribe personality to our cars, appliances, and other inanimate objects. Consider the familiarity of such concepts as *temperamental thermostat, stubborn lock,* and *finicky laptop.*

Behavior chains, like their constituent reflexes, can be used to define and support the intended personality of an android. For example, if you're designing a MEC soldier, then the appropriate reflex response to a flash of light or loud sound is an offensive posture, ready for attack. A high dominance score in the profile of the MEC soldier reflects the expected behavior.

If your goal is to design a believable android, then consider building in some randomness in the behavior chains. You could also design the sensor-system response so that not only is the JND a constant proportion of the initial sensory level, but also the absolute sensitivity is within normal human range. The temperature sensors on an android's skin or exoskeleton might be able to detect a 0.01 percent change in temperature, but it had best not appear to act on that data if the goal is to produce a human-like android.

Behavior-Chain Experiments

Armed with simple reflex arcs patterned after those described in Chapter 1, a robot arm, temperature and pressure sensors, and a little imagination, we'll explore a real-world example of the theoretical behavior chains outlined in Figure 2-1. Short of a full-fledged android, a robot arm is a good experimental platform because it's complex enough to demonstrate a range of physical behavior chains. Another option is to use simulation. However, computer simulations may provide insight into what should happen, but there's something magical about moving real atoms in the real world.

As shown in Figure 2-3, we'll start things off with a press of S1 to initiate closure of the arm's gripper. If the temperature of the object is excessive, as measured by a

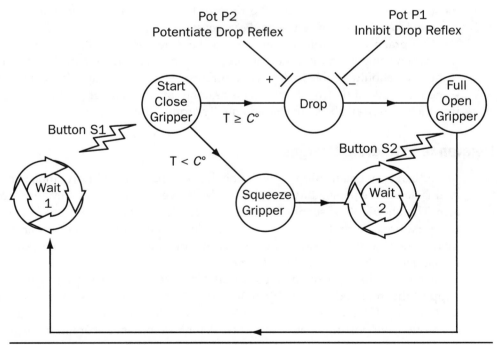

FIGURE 2-3   Overview of our behavior-chain experiment.

sensor on the gripper, the gripper will open fully to drop the object. After a brief refractory period, in which the temperature sensor is allowed to cool to room temperature, the system will again respond to the closure of S1. We'll simulate the excitatory effect of adrenaline and the inhibitory effect of the brain on the drop reflex with two potentiometers, P1 and P2, respectively.

If, however, after the gripper makes contact with the object, the temperature is not excessive, then the gripper will squeeze (i.e., hold) the object, as measured by a pressure sensor on the gripper. During this time, the object can be moved from one place to another or simply held motionless in space by the robot arm. Pressing S2 opens the gripper fully, enabling the system to respond to the press of S1.

Bill of Materials

Our platform for exploring behavior chains is a 5-degree-of-freedom (DOF) robot arm built with low-cost analog hobby servos. The major advantages of using a robot arm built with traditional servos include ease of use, quick setup, and a wide range of compatible hardware and software. For example, there is an Arduino library for the Sony PS/2 controller and Arduino-compatible cards such as the BotBoarduino that make working with multiple servos a breeze. As long as you use analog hobby servos, you'll be able to use the code examples contained herein and the standard Arduino libraries.

Another option is to use an arm built with AX-12A Dynamixel servos. Although these digital servos are more difficult to work with because of their odd mounting hardware and non-Arduino-compatible serial interface, they have built-in sensors for temperature, position, load voltage, input voltage, and compliance of the output shaft. There's an excellent Dynamixel servo library for the Arduino at Savage Electronics worth exploring—as long as you're willing to add a chip to the serial control line. The sensor handling routines offered in this chapter should be easy to retrofit into a Dynamixel or other digital servo environment, but control of the servos proper necessarily will deviate from what's described here.

Don't have a few hundred dollars to spring on a robot arm? Then consider buying or building a single-servo gripper without the arm. Alternatively, you can follow along with the two-servo platform we used for the simple reflex arc simulator in Chapter 1. You can put indicators on the servo horns to help you visualize bending of joints and closure of the gripper.

In addition to the robot-arm platform, Arduino microcontroller, and Sony PS/2 controller or other means to interactively direct arm position, you'll need the following:

- Power resistor or other heat source
- Power supply for heat source (optional, depending on source)
- TMP36 analog temperature sensor
- 10-kΩ thermistor

- Cadmium-sulfur (CdS) photocell
- Force-sensing resistor
- Momentary contact button switches (2)
- 10-kΩ potentiometers (2)
- 5-V direct-current (dc) power supply for Arduino and servos
- Jumpers or wires
- Shield of your choice (optional)

The components used for this series of experiments, shown in Figure 2-4, include a standard Arduino Uno, a Grove shield, Grove buttons and potentiometerss, force and temperature sensors, and a Lynxmotion robot arm. Note the power resistor used for a heat source.

Nice to have, but not necessary, is a calibrated temperature probe and digital multimeter (DMM) to verify the temperature of the heat source. Fingers are a good alternative to a heat source, albeit they're not as impressive as a cup of coffee or a massive power resistor. Not shown in the figure are the power supplies for the power resistor, Arduino, and robot arm and the numerous cables to connect the robot-arm servos, sensors, and buttons and potentiometers to the Grove shield.

FIGURE 2-4   Components of the authors' experimental setup.

The heat source is up to you. The choice probably will come down to the size and lift capability of your robotic gripper. Fortunately, even the grippers on a $39 toy robot arm can handle the size and weight of a 5-W power resistor and short power cables.

Circuit

Figure 2-5 shows the overall circuit for the first experiment. Mechanical and electrical details for the elbow, shoulder, and base servos, along the right side of the figure, are not shown for clarity. Electrically, each servo has a 5-Vdc source and ground from the power supply and a pulse-width-modulated (PWM) connection from the Arduino. The mechanical gripper device, with embedded temperature and force sensors, is shown in the lower-left corner of the figure.

The force-sensing resistor FSR1 isn't mandatory. If you're good at operating the gripper of your robot, you can leave out that component. However, if you're not that facile with stopping the jaws of the gripper at just the right moment, you risk stripping the gears of the gripper servo. In this regard, FSR1 is relatively inexpensive insurance. Besides, it'll come in handy with other experiments with your robot arm.

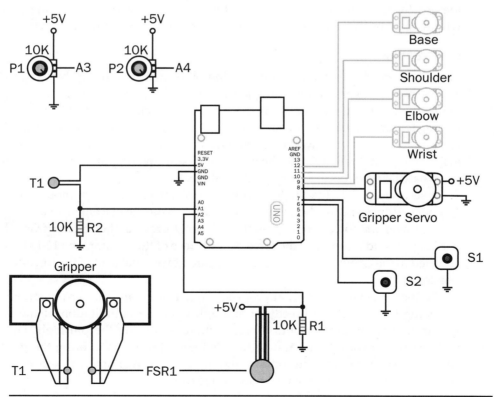

FIGURE 2-5   Schematic of the behavior-chain experiment.

The 10-kΩ thermistor T1 is used with a 10-kΩ resistor R2 to form a voltage divider across the 5-Vdc supply line. Thermistors exhibit a negative temperature coefficient—increasing the temperature decreases the resistance of the thermistor. In our circuit, which uses a 10-kΩ thermistor from Sparkfun, the ambient temperature reading from the Arduino is about 450 (2.19Vdc). Gripping a thumb increases the reading to about 475 (2.32Vdc), and the response is rapid—less than a second.

Also, the reading *increases* with temperature because of how the voltage divider is configured. The lower the resistance of T1, the greater is the voltage drop across R2, and the greater is the voltage applied to the analog input of the Arduino.

Because resistance change in T1 with temperature is nonlinear, you'd need a calibrated thermometer or temperature sensor to collect the data for a lookup table or equation to translate the voltage drop across the thermistor to an accurate temperature reading. You could also adjust the lookup table or equation to reflect the self-heating of the thermistor. Because thermistors present a relatively low resistance of 10 kΩ or so compared with 1 MΩ or more for solid-state sensors, current and the resulting heat dissipation can be significant. Fortunately, this overhead isn't necessary because we're just after significant changes in relative temperature for now. Moreover, thermistors are inexpensive and respond rapidly—within a few hundred milliseconds—to changes in temperature.

Construction

There isn't much involved in construction other than mounting the temperature and pressure sensors on opposing jaws of the gripper. Use double-sided tape to mount Grove terminal blocks on the outside of each jaw. One block holds the thermistor and 10-kΩ resistor, shown mounted in Figure 2-6. Whether you use the terminal block or your own board, mount the thermistor in adhesive foam on the inside of one gripper jaw, with the insulating foam covering the otherwise exposed leads between the thermistor and terminal block.

A second block holds the force-sensing resistor FSR1, shown mounted on the gripper jaw opposite the thermistor in Figure 2-7. Note the placement of an adhesive footpad over the force-sensing resistor. The pad protects the resistor from sharp, hard objects and distributes the force over the face of the resistor. The 10-kΩ resistor attached to the Grove terminal block is obscured by the flat lead of the force-sensing resistor.

Connect the switches to the appropriate digital inputs and the two potentiometers to analog inputs, either directly or through a shield. A limitation of the Grove system is that the default button-press action is to bring the pin high. This is a problem if you want to use the Arduino's internal resistors to bring the pin high and then use an external button to bring the pin low. A simple work-around is to swap the ground and Vcc wires in the Grove cable.

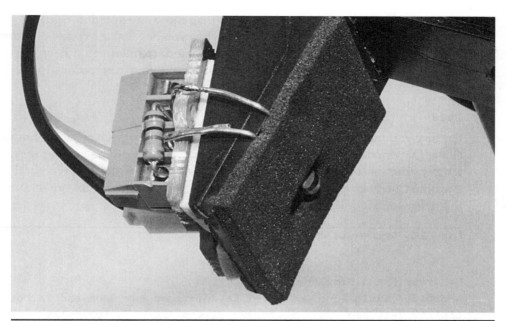

FIGURE 2-6   Thermistor mounted in adhesive foam on the gripper jaw. A 10-kΩ resistor is also attached to the Grove terminal block.

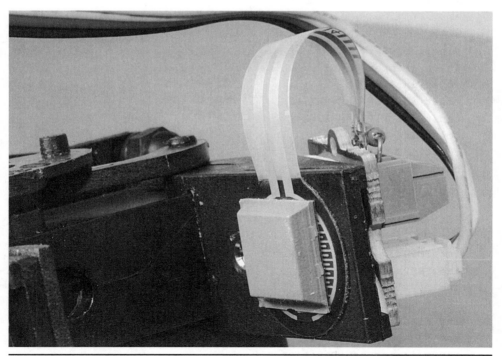

FIGURE 2-7   Force-sensing resistor and 10-kΩ resistor in a Grove terminal block mounted on the gripper jaw.

If you have to use Arduino pins other than those suggested in Figure 2-5, then remember to change the pin assignments when you develop the code. For example, if you're using the BotBoarduino with PS/2 interface, then you'll have to shuffle the pin assignments around to make room for the four connections to the PS/2 controller.

Code

Because you may not have an arm with the same electromechanical configuration as the Lynxmotion arm used in our example, we focus on the gripper servo and temperature sensor. In addition, we'll use the standard Arduino servo library and Arduino Uno to keep things simple. You shouldn't have any trouble integrating the following code into your specific robot arm control code.

LISTING 2-1   Arduino code for the behavior-chain example.

```
/*
Behavior Chain Simulator
Androids: Build Your Own Lifelike Robots by Bergeron and Talbot
Listing 2-1. See www.mhprofessional.com/Androids for fully
documented code
Arduino 1.0.1 environment
*/

#include <Servo.h>
Servo gripperServo;
const int TMPPin = A1;
const int FSRPin = A2;
const int P1Pin = A3;
const int P2Pin = A4;
const int S1Pin = 7;
const int S2Pin = 6;
const int gripperServopin = 8;

int TMPValue = 0;
int TMPValueStartup = 0;
int TMPDifference = 10;
int FSRValue = 0;
int P1Value = 0;
int P2Value = 0;
int S1ButtonState = 1;
int S2ButtonState = 1;
int grippperServoPosition = 0;
int noiseLevel = 2;
```

```
int directionFlag = 0;
int servoSettleDelay = 5;

void setup() {
gripperServo.attach(gripperServopin);
  pinMode(S1Pin,INPUT);
pinMode(S2Pin,INPUT);
digitalWrite(S1Pin,HIGH);
digitalWrite(S2Pin,HIGH);
gripperServo.writeMicroseconds(1000);
TMPValueStartup = analogRead(TMPPin);
}

void loop(){
  P1Value = analogRead(P1Pin);
P1Value = map(P1Value, 0, 1023, 0, 100);
P2Value = analogRead(P2Pin);
P2Value = map(P2Value, 0, 1023, 0, TMPDifference);
TMPValue = analogRead(TMPPin)-TMPValueStartup;
S1ButtonState = digitalRead(S1Pin);
 if (S1ButtonState == LOW){
  if (TMPValue < TMPDifference){
    FSRValue = analogRead(FSRPin);
     if (FSRValue < noiseLevel){
    grippperServoPosition = grippperServoPosition + 1;
    grippperServoPosition = min (grippperServoPosition, 180);
    gripperServo.write(grippperServoPosition);
    delay(servoSettleDelay);
    }
}
  }
S2ButtonState = digitalRead(S2Pin);
 if (S2ButtonState == LOW){
   dropit();
  }
if (TMPValue >= (TMPDifference + P1Value - P2Value)){
   dropit();
  }
 }

/*
  -----------------------------------------------------------
```

```
    dropit()
    ------------------------------------------------------------
    */
```

```
void dropit() {
grippperServoPosition = 0;
gripperServo.write(grippperServoPosition);
}
```

Now, working from top to bottom of Listing 2-1, we first load the servo library and then define the pinouts on the Arduino Uno and initialize the variables in the program. At startup, we measure and store a value representing the ambient room temperature in `TMPValueStartup`.

Within the main loop, we read the values of P1 (inhibitor) and P2 (exciter) for the drop reflex and map these onto `P1Value` and `P2Value`. The value from P1 is mapped onto 0–100, whereas that of P2 is mapped onto `0-TMPDifference`, the difference in temperature between the initial ambient and current thermistor reading set as the trigger value. In this example, it's only 10 because we're dealing with relatively low differences in temperature.

When button S1 is pressed, the software first checks the difference in current thermistor reading with that of the established `TMPDifference`. If the temperature difference is low enough (i.e., the thermistor has cooled sufficiently) and the force-sensing resistor isn't measuring pressure above threshold, then—and only then—is the gripper closed. S2 is the manual override. Hit S2 and the jaws open immediately.

The deceptively simple line below is responsible for the semiautonomous behavior directed by "cortical" inhibition (`P1Value`) and "adrenal" excitation (`P2Value`):

```
if (TMPValue >= (TMPDifference + P1Value - P2Value)){ dropit();}
```

As P1 is turned clockwise, `P1Value` increases, raising the effective temperature difference in the object versus the ambient temperature required for the object to be dropped. Conversely, as P2 is turned clockwise, `P2Value` increases, diminishing the effective temperature difference in the object versus ambient temperature required for the object to be dropped. In this example, `P1Value` ranges from 0–100, whereas `P2Value` ranges from `0-TMPDifference`. In this way, the gripper is inhibited from constantly dropping the object regardless of the relative temperature difference.

Operation

In operation, use your Sony PS/2 controller, joystick, bank of momentary switches, or other control setup to move the base, shoulder, and elbow joints to position the wrist to pick up the potentially hot object. Then press and hold momentary contact button S1 until the jaws make contact with the object. Then use your control circuitry to move the other servos to pick up and move the object. Of course, if you're moving a resistor with power leads attached, then be careful not to dislodge the power connection. You can use alligator clips to deliver power to the resistor or use 3.5-mm bullet connectors on the ends of the resistor and on the power cable.

Another option is to use your thumb as the "hot object," as in Figure 2-8. Just make certain that the fleshy part of your thumb faces the thermistor. Fingernails are relatively good insulators and make a much better target for the force-sensing resistor than the soft tissue of the thumb.

As outlined earlier in Figure 2-3, if the object is below the trip temperature established by the software, then the gripper and arm will work as you direct it. That is, when you've positioned the resistor or other object where you want it, pressing button S2 will release it. The gripper jaws will then open fully, and the wrist will return to the ready position. Use your other controllers to move the base and other joints to their original positions for another test run.

However, if the temperature of the object is at or above the trip temperature that you defined, and P1 and P2 are fully counterclockwise, then the gripper will open, dropping the object. After a few seconds, when the thermistor has cooled sufficiently, the gripper will again respond to the press of S1.

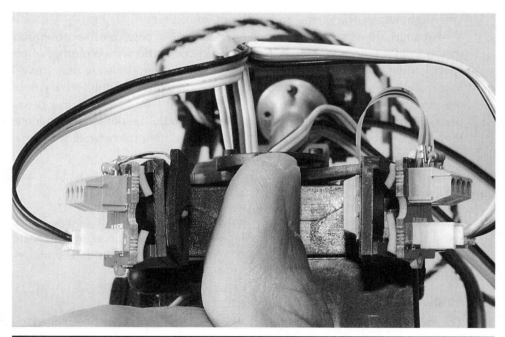

Figure 2-8   Robot gripper poised to clamp down on a warm thumb.

As outlined earlier, if you turn P1 clockwise—that is, increase the resistance provided by P1, the autonomous dropping behavior would be inhibited. At the extreme clockwise position of P1, there may be no autonomous dropping behavior at all depending on the temperature of the object. Think hot cup of coffee held over your head. You wouldn't want your service android to drop the coffee regardless of the damage it might inflict on itself. Ideally, your android will have some higher-order pattern-recognition device that recognizes your impending peril and effectively maximizes the value presented by P1.

Rotating P2 clockwise negates the effect of P1, but only to a point, because of the relative scaling of the two pots. However, with P1 fully counterclockwise, P2 can excite the response to the point that noise in the thermistor can set off the drop reflex. All this interaction, of course, depends on the value assigned `TMPDifference` and the relative weights given P1 and P2.

This simple behavior chain with three linked reflex arcs has practical value. Imagine, for example, that your service android's hand is covered with expensive, lifelike latex. If you instruct your android to pick up an unexpectedly hot pot from the stovetop, then, assuming that you've provided it with this behavior chain, you won't have to worry about repairing burned latex.

The responsiveness of a temperature-sensing gripper to human skin and muscle also suggests a safety application. Imagine an assistant-cook android grasping for a tomato to prep for a salad when you inadvertently grab for the tomato at the same instant. With the temperature-sensing and instant-release behavior installed, you might escape the slicing and dicing of your digits.

An alternative to this simple behavior chain is to consider the consequences of the android's every movement in, say, the kitchen, and write volumes of context-specific rules for the central processing unit. However, odds are that you'd forget to include rules of unlikely but possible situations, such as "don't stick you head in the oven when checking on the pies." For some situations, it's much easier to provide simple, low-level behavior chains to handle the potential accidents, with the equivalent of cortical inhibition and adrenal excitation when needed. Remember the mice and the maze.

Modifications

The challenge of working with any robot-arm setup is holding back. It's hard to resist the urge to add sensors for color, sound, vibration, and so on because the platform is so flexible and expressive. Following is a sample of the many mods worth considering.

Absolute Temperature Reference

The thermistor T1 is good for quick relative temperature measurements. However, you may need accurate absolute temperature measurement that can be used to

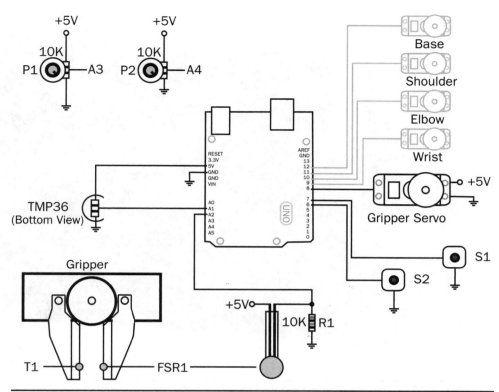

FIGURE 2-9 Schematic showing solid-state temperature detector TMP36.

trigger a behavior chain that fires when the temperature is within, say, 2 degrees of a preset limit. This can be accomplished by creating a temperature lookup table for a thermistor. A simpler approach is to replace thermistor T1 with a solid-state absolute temperature sensor such as the TMP36. Figure 2-9 shows the updated schematic with the TMP36 in place.

Setup of theTMP36, which looks like an ordinary three-terminal epoxy transistor, couldn't be simpler—one terminal goes to +5 V, one goes to ground, and the center terminal goes to analog input A1. Figure 2-10 shows the TMP36 mounted in one jaw of the gripper. Note that the flat face of the TMP36 is against the jaw for stability. Alternatively, the flat face can be oriented inward to maximize surface contact area and therefore heat transfer from the "hot" object. If you opt for this orientation, you'll have to fashion something in the jaw to prevent the TMP36 from rolling if the "hot" object isn't perfectly centered on the flat surface of the device.

The TMP36 is essentially a temperature-dependent potentiometer, in that changes in temperature are analogous to changing the position of the wiper terminal on a potentiometer. The output voltage of the center terminal is directly

FIGURE 2-10   TMP36 temperature sensor mounted to one jaw of the gripper.

proportional to the case temperature, at 10 mV/°C, according to the following formula:

$$\text{Temperature (°C)} = [V_{out}\,(\text{mV}) - 500]/10$$

At room temperature (25°C), with 5.0 V applied to the sensor, the output is about 750 mV. By the way, when you're testing the TMP36, a great thing to have around is a can of compressed air. You can cool down the TMP36 to 0°C from 30°C with a 2-second blast of compressed air—invert the spray can as you release the air to maximize the rate of cooling.

You'll need to modify the code of the preceding thermistor-based example only slightly to accommodate the TMP36. The modified code appears in Listing 2-2.

LISTING 2-2   Arduino code for behavior chain with TMP36 temperature sensor.

```
/*
Behavior Chain Simulator
Androids: Build Your Own Lifelike Robots by Bergeron and Talbot
```

```
Listing 2-2. See www.mhprofessional.com/Androids for fully
documented code
Arduino 1.0.1 environment
*/

#include <Servo.h>
Servo gripperServo;

const int TMPPin = A1;
const int FSRPin = A2;
const int P1Pin = A3;
const int P2Pin = A4;
const int S1Pin = 7;
const int S2Pin = 6;
const int gripperServopin = 8;

float TargetTMPC = 30.0;
float CurrentTMPC = 0.0;
int TMPFlag = 0;
int FSRValue = 0;
int P1Value = 0;
int P2Value = 0;
int S1ButtonState = 1;
int S2ButtonState = 1;
int grippperServoPosition = 0;
int noiseLevel = 2;
int directionFlag = 0;
int servoSettleDelay = 5;

void setup() {
  gripperServo.attach(gripperServopin);
  pinMode(S1Pin,INPUT);
  pinMode(S2Pin,INPUT);
  digitalWrite(S1Pin,HIGH);
  digitalWrite(S2Pin,HIGH);
  gripperServo.writeMicroseconds(1000);
}

void loop(){
  P1Value = analogRead(P1Pin);
  P1Value = map(P1Value, 0, 1023, 0, 100);
  P2Value = analogRead(P2Pin);
```

```
    P2Value = map(P2Value, 0, 1023, 0, 10);
  S1ButtonState = digitalRead(S1Pin);
  if (S1ButtonState == LOW){
     TMPCheck();
     if (TMPFlag == LOW) {
     FSRValue = analogRead(FSRPin);
     if (FSRValue < noiseLevel)  {
     grippperServoPosition = grippperServoPosition + 1;
     grippperServoPosition = min (grippperServoPosition, 180);
     gripperServo.write(grippperServoPosition);
     delay(servoSettleDelay);
     }
   }
   }
 S2ButtonState = digitalRead(S2Pin);
  if (S2ButtonState == LOW){
    dropit();
   }
     TMPCheck();
     if (TMPFlag == HIGH) {
     dropit();
   }
   }


   /*
 ----------------------------------------------------------
 dropit()
 TMPCheck()
 ----------------------------------------------------------
 */

 void dropit() {
    grippperServoPosition = 0;
    gripperServo.write(grippperServoPosition);
    }

 void TMPCheck() {
    int reading = analogRead(TMPPin);
    float voltage =  reading * 5.0;
    voltage /= 1024.0;
    CurrentTMPC = (voltage - 0.5) * 100;
   if (CurrentTMPC >= TargetTMPC) {
```

```
    TMPFlag = 1;
  }
    else {
    TMPFlag = 0;
    }
  }
```

Working from the top down, the first major change in the program is the addition of two variables `TargetTMPC` and `CurrentTMPC`, both declared as type float. `TargetTMPC` is the target temperature, in Celsius, at which the object should be dropped. As with the preceding thermistor-based version, the effective drop temperature can be altered by P1 or P2, representing "cortical" inhibition and "adrenal" excitation, respectively.

Instead of establishing a baseline temperature at startup, the current temperature is repeatedly measured and compared with the target temperature. The function `TMPCheck( )`, or temperature check, first performs the conversion of the sensor output voltage to temperature and then performs the comparison of current and target temperatures. The result of the comparison is used to set the value of `TMPFlag`, or temperature flag, which is high when the target temperature has been reached.

The TMP36 is more temperamental than the 10-kΩ thermistor used in the preceding listing. It occasionally outputs a spurious high or low value. The specification sheet for the TMP36 calls for a 0.1-mF ceramic disk capacitor from the +5-V source to ground terminal of the device, but this had little effect on the spurious values. And we tried several TMP36 devices, all of which behaved identically. The one thing that did help was adding a delay of 5 ms or more after each read of the device. The downside of this fix is that it slows the rate of closure of the gripper somewhat.

Minor annoyances aside, the ability to accurately detect the temperature of the object in the grips of an android can be leveraged in any number of situations. In addition to the safety issues discussed earlier, think of the value of a home-nurse robot that can accurately measure skin temperature with a touch or that can determine, within a few tenths of a degree, whether your bath water is at your preferred temperature.

Night Owls

Want to create a nocturnal android with behavior links that are faster when it's dark than when it's light out? Substitute a cadmium-sulfur (CdS) light sensor and 10-kΩ voltage-divider resistor for P2, as in Figure 2-11. Mount the sensor, CdS1, in an opaque container so that you can easily control the intensity of the light. Opening and closing the lid on a small container is easier and less disruptive than turning the room lights on and off during your experiments. Although a thermistor is

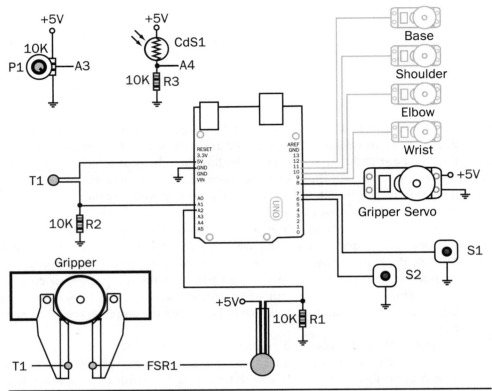

FIGURE 2-11 Schematic showing substitution of a CdS light sensor, CdS1, for pot P2 to enable illumination-mediated behavior.

shown with this modification, it can be used with either the thermistor T1 or the solid-state temperature sensor TMP36.

The CdS sensor presents around 300 kΩ in the dark and only about 10 Ω when brightly illuminated. As such, the voltage across the voltage-divider resistor R3 is greatest when the CdS detector is illuminated. You should be able to leave the code framework as is and modify the intensity of illumination to alter the excitatory effect on the drop behavior. You will likely have to modify the mapping function for `P2Value`:

```
P2Value = map(P2Value, 0, 1023, 0, 10);
```

Instead of mapping the input range to `1-10`, a rather course granularity, try mapping the CdS voltage to `0-50` or `0-100`. Depending on your light source and ambient light intensity, you should find that mapping to a larger range provides more sensitivity to small changes in light intensity.

Figure 2-12 Grove light-sensor module with CdS sensor in the lower right.

If you're using the Grove shield, then an easy way to add a CdS sensor is to use a Grove light-sensor module, shown in Figure 2-12. In addition to a standard CdS sensor, the module has a built-in resistor and active buffer so that you can read the module directly without an external resistor.

Once you get the circuit working with massive swings in light intensity—as in opening and closing an opaque box containing the CdS sensor—you should try detecting subtle changes in illumination. Here's where the concept of just noticeable difference (JND) comes in. Your challenge is to adjust the weighting of the CdS output so that a change in illumination that is just perceptible to you 50 percent of the time results in a comparable change in CdS output.

For this experiment, you can use a dark room and a monochromatic light source such as a red light-emitting diode (LED) and a light intensity meter—either an add-on for your DMM or a stand-alone model sold for photography. One of the challenges of mapping your JND to the output of the CdS sensor is that the

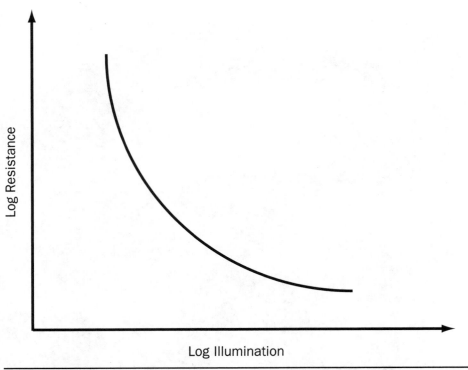

FIGURE 2-13   Nonlinear relationship of CdS sensor resistance versus illumination.

relationship between resistance and illumination is nonlinear, and the nonlinearity depends on the absolute level of illumination, as in Figure 2-13.

Of course, if you start changing the color of light with illumination level—common with artificial lighting—then you have another variable to consider. The human eye, like a CdS cell, responds differently to light of different wavelengths. So use a red, green, or blue LED to minimize the spectral variance with illumination intensity.

Remember, the goal is to create a map or table for the CdS sensor that mimics the JND of the human eye. If this task seems a little daunting at the moment, then you can revisit this challenge after reading about light sensors and the eye in Chapter 4.

Attitude Adjustment

This mod requires a bit of imagination on your part. Using the personality-profile indices, devise postures and response rates that correspond to a dominant, aggressive personality, that is, a posture during and after the manipulation of the object that scores high for dominance, negative for conscientiousness, neutral to negative on stability, and extremely negative for openness. Just how do we realize this with a few blocks of plastic stuffed with motors and electronics?

It's easier than you might imagine. Move past the arm, and think of the dominant, aggressive person that could be attached to the arm. If that person goes to pick up an object and drops it out of reflex, she's not going to whimper "Ouch" and curl up in a ball. No, she might yell out a profanity or two (negative conscientiousness) and instinctively ball her hand into a fist ready to strike (dominance) and then immediately pick up the object (questionable to negative stability) despite the pain.

Let's pass on the profanity for the moment and work on the other behavior chains. We need to replace the simple "withdraw-delay-return" behavior with "draw-return" behavior. In addition, we need to withdraw to an aggressive posture, as opposed to simply moving away from the hot object.

Using your controller and arm-control libraries, define a position for your arm that looks aggressive—for example, gripper forward, coiled, prepared to strike, as a snake. When you're done, do the same for neutral and submissive personalities. Figure 2-14 shows our take on the postures.

This sort of setup is easily done with arms composed of Dynamixel servos because you can move the arm to a position that you like and then query the servos for their position. There's also a pose application for the Lynxmotion arm that serves the same purpose.

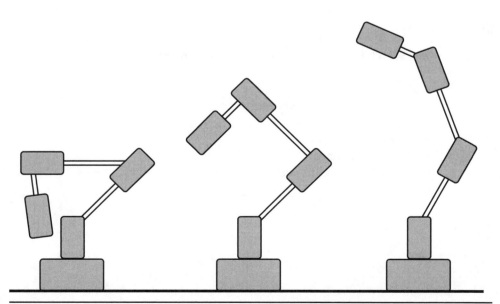

FIGURE 2-14 Posture personality configurations, from left to right: submissive, neutral, and dominant.

Gremlins

This chapter is fairly straightforward, assuming that you have a debugged robot arm as your platform. Even so, hobby servos are often problematic and interfere with the operation of other servos and attached electronics. If you get inconsistent behaviors, make certain that your servo and Arduino power supplies are either separate or highly filtered. Another problematic area associated with servos is dead zones—positions where the motors provide diminished power. Your robot-arm library should handle the dead zones, as well as other reverse-kinematics issues inherent in servo-controlled arms.

For example, assume that the goal is to move from the current pose to a pose in which the elbow joint is at a right angle, perpendicular to the ground. With manual control of a robot arm, it's possible to extend the arm fully and then bend at the shoulder. However, most servos won't stand this kind of abuse, and you'll either strip the gears or burn out the motor in the shoulder servo. A better way is to bend both shoulder and elbow joints simultaneously, avoiding placing the entire load on one servo. This is another reason to keep your experimental setup to a gripper and wrist servo if you don't have a full robot arm with library.

There's also the issue of safety. If you use the higher-torque analog servos or the digital Dynamixel servos, then be careful around pinch points. You can get a nasty skin tear if you're not careful. Think of each joint as a motorized pair of scissors— not something you'd want to run with, even when powered down.

Search Terms

Relevant search terms for your Web browser include

- Behavior chains
- Emotionally intelligent interfaces
- Thermistor
- Reverse kinematics
- Servo dead zones
- Just noticeable difference
- Weber constant

CHAPTER 3

Homeostasis

Homeostasis—the maintenance of constant internal conditions—is the purpose of much of our physiology. Moreover, the physiology that enables us to stand rapidly from a seated position without passing out has direct parallels with the proportional-integral-derivative (PID) controller that enables a robot drone to maintain level flight despite windy conditions. Both rely on negative-feedback control-loop mechanisms.

In this chapter we explore homeostasis and basic control theory using temperature regulation as a focus. We chose temperature instead of, say, joint velocity, power, or pressure because of readily available sensors, universal firsthand experience, and perhaps most important, the rare chance to intentionally mix water and electricity. There's also the direct applicability to practical android designs. Not only are electronics designed to operate within a limited range of temperatures, but an android with a cold handshake is certain to make a bad first impression.

We'll create a model of skin-based cooling using a water-filled heat exchanger made with a few feet of silicone tubing and then model the body core with an insulated container, a 12-V direct-current (dc) water pump, a pair of thermistors, and a power metal-oxide semiconductor field-effect transistor (MOSFET). Toss in an Arduino and a PID algorithm, and we're good to go.

Biological Basis

For humans, homeostasis is about staying alive. Within the body at any given moment, there are hundreds of ions, thousands of proteins, and tens of thousands of other biologically active molecules in flux. On a macro level, this apparent chaos is observable as changes in, for example, heart rate, breathing rate, skin temperature, and relative blood flow to the skin and organs. Temperature regulation is a top-tier contributor to the homeostatic system. Without it, just about every other process breaks down.

65

Dark Side of the Moon

Sci-fi movies commonly place the hero stranded in a space suit on some uninhabited ship or moon with only a few hours of air remaining. And there's often some dialogue about whether it's better to die from lack of oxygen or from carbon dioxide poisoning. More realistically, though, the intrepid explorer is going to die from hypothermia long before the air tanks or carbon dioxide scrubbers fail. It takes a lot of energy to keep a suit warm when the ambient temperature is a few hundred degrees below zero.

So what can our hero expect as the suit's heating system fails? Initially, shivering, accompanied by rapid breathing and heart rate. There's commonly an intense need to pee as well as mental confusion—not necessarily a great combination in a space suit depending on how it's equipped.

As the body cools, the shivering becomes more violent, and it's difficult to perform fine-motor tasks, like turning the valve on a newly discovered air tank. Walking becomes difficult. Warm blood is shunted away from the periphery, leaving the lips, ears, fingers, and toes blue, cold, and unresponsive. This is a good time for the rescue ship to appear, just before it's too late to reverse the process.

Barring a last-minute rescue, our explorer's body starts shutting down with a further drop in body temperature. Breathing becomes slow and labored. The heart slows, and beating becomes irregular, resulting in an abnormally low blood pressure. The organs, including the brain, start failing. The astronaut may try tearing off the suit or looking for a nook to curl up in a fetal position.

Clearly, this is not a pleasant way to go. However, the scenario outlines some of the mechanisms used by the body to maintain thermal homeostasis. First, there's increased physical activity—shivering, increased heart rate, and increased breathing rate. Metabolism kicks into high gear. With decreasing core body temperature, the intensity of heat-producing shivering increases, and warm blood is sequestered in the core, away from the cold. After this point, it's simply a matter of system breakdown. Organs fail, and death is inevitable.

Temperature Regulation

With this "heart-warming" tale as a backdrop, let's review the basics. For the average adult sitting in a typical doctor's office with an ambient temperature of 72°F and the unpleasant smell of disinfectant in the air, the core body temperature is about 98.4 ± 1°F (37.0 ± 0.6°C). The core temperature increases a degree or two beyond this average after meals, physical activity, and fever. A quick dip in the pool can lower the temperature by a degree or so. Core temperature tends to peak in the early afternoon and dip just before waking.

Core temperature is, as the name suggests, the temperature of the center or core of the body as opposed to the temperature of the fingertips or earlobes. Moving from the core to the periphery, body temperature drops as a function of age, height, activity level, and gender. For example, physiologically correct female androids have cold feet regardless of ambient temperature.

Plunge a thermometer directly into the middle of someone's brain or liver, and you're measuring core temperature—which is why temperature measurements at these sites are used to determine time of death. If you know the ambient temperature and the core temperature, you can determine rate of temperature drop and therefore time of death, assuming a cool ambient temperature and a relatively recent death. Oral and rectal temperature measurements are less dramatic but also less precise measures of core temperature.

Temperature regulation is key to human survival because our biological processes have evolved to operate within a fairly narrow temperature range. Muscle contraction and relaxation, for example, don't work well in the cold. You might survive a plane crash into the ocean, but if there are sheets of ice floating about, then your heart will stop beating after only a few minutes as the muscle fibers lose their elasticity and your blood thickens. At least you won't feel the shark bites.

On the other hand, if you're stranded outside in a blizzard without a jacket, ice crystals will form in the cells of your brain, eyes, muscles, and skin, rendering you blind and immobile. This would be a great time to go into shock, especially if there is a pack of wolves downwind. Ice crystal formation within cells—such as occurs in frostbite—is a reason previously frozen meat doesn't taste as good as fresh meat and why cryogenic preservation for intergalactic travel is problematic.

A high fever can be just as deadly as being locked in a car that's parked in the sun on a hot day without a crack in the window. If you've ever cooked an egg in aluminum foil on your car's dashboard—the "dashboard deli"—you know that it doesn't take much heat to cook an egg or make a cheese melt. Proteins denature—that is, break down—and become inoperable at temperatures above about 106°F (41°C). The denaturing process is reversible for some proteins, but for many others, such as the albumin in egg whites, well, there's no turning back. Permanent denaturation of proteins is the reason you might survive a high fever and later discover that you've lost thousands of neurons and a few IQ points in the process.

It turns out that we have about 10 times as many cold sensors in our skin as heat sensors. Conversely, the temperature sensor in our brain—part of the hypothalamus—is disproportionately sensitive to the heat of the core. From a survival standpoint, this makes sense. Exposed to the cold, our primary homeostatic mechanisms for maintaining core body temperature are shivering and shunting blood away from the skin and periphery.

The problem with shunting blood away from the surface is that while this can temporarily keep the core temperature from dropping, the hands and feet become susceptible to frostbite. Frostbite may seem inconsequential today, but without medical treatment, a hand or foot with frostbite at best becomes a stub. At worst, the rotting, gangrenous tissue becomes infected and bloated. In addition to being unsightly and malodorous, the dead and decomposing tissue breeds bacteria that spread into the bloodstream, causing death.

Excess heat is less problematic to the extremities. Most of our heat loss at rest occurs through radiation and conduction. Removal of heat by convection can be significant if there's a nice breeze blowing. These three mechanisms of heat removal work as long as the environment is cooler than the body, but what if it's 110°F in the shade?

This is where sweating comes in. A part of the hypothalamus, sensing the increased core temperature, turns on the evaporative cooling system. Not only does the skin open the sweat floodgates, but the skin blood vessels near the surface dilate, shunting the hot blood from internal organs to vessels just beneath the surface of the skin, resulting in rapid heat loss. Shivering and other heat-generating mechanisms are also inhibited.

Our centrally mediated sweat response to a rise in core temperature is imperfect. It doesn't work in a high-humidity environment, for example. In addition, kicking out two to three liters of sweat per hour is fine as long as you have a supply of cool water, but once you're dehydrated, the sweat mechanism shuts down. Peripheral temperature rises to ambient, and once core temperature hits the magical 106°F, the proteins in your brain might as well be egg whites in a hot frying pan.

Negative Feedback

The temperature-regulation system, as with most control systems in the body, is based on negative feedback. If you're familiar with op amps and phase-locked loops (PLLs), then Figure 3-1 will look familiar. As shown in the figure, the metabolic activity in the body, which results in heat generation, affects core body temperature.

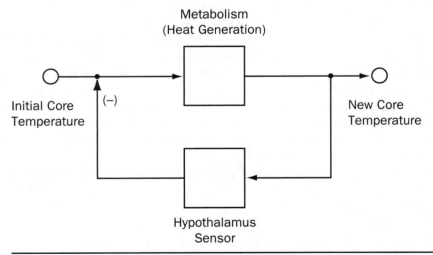

FIGURE 3-1   Block diagram of a negative-feedback temperature-control system. Part of the hypothalamus serves as the core body temperature sensor.

Sensors in the hypothalamus respond to a rise in core temperature above the set point by generating negative feedback on the system to bring the temperature back to the set point. Sweating, increased blood circulation in the skin, decreased blood circulation in the muscles and organs, and perhaps panting with your tongue hanging out are all negative-feedback responses that dissipate heat and reduce core temperature.

From Figure 3-1 you can see that metabolism increases core temperature, activating the hypothalamic temperature sensor, which negatively regulates metabolism. For example, when the hypothalamus inhibits shivering, core temperature falls.

Gain is a useful concept in discussing a negative-feedback system because it's a measure of how well the system maintains homeostasis. For thermal homeostasis, gain can be expressed as

$$\text{Gain} = (\text{change in environmental temperature}/ \text{change in core body temperature}) - 1$$

For example, assuming that a reasonable 25°F change in the environment results in a 1°F change in core body temperature, then the gain of the system is

$$(25°/1°) - 1 = 24$$

The set point established by the hypothalamus isn't hardwired to 98.4 or 98.6°F but can be affected by variables such as the skin temperature. If core is cooler than normal, but the skin temperature sensors indicate a relatively high ambient temperature, then the set point for shivering is depressed. Conversely, if the skin sensors indicate low temperature, then the set point for shivering is raised. In a way, the skin temperature sensors constitute an early warning system that responds to the environment minutes before the core. For example, if you jump into a nice, cool stream to cool off, it would be a waste of energy to start sweating, even if your core temperature were a little elevated. Similarly, it would be a waste of calories to start shivering once you've arrived at a warm home. In this way, the skin sensors are predictors of future core temperature.

It's important to note the behavioral aspects of homeostasis, such as seeking shelter from the cold or jumping into a cool body of water to reduce body temperature. Behavior such as putting on a heavy coat can override the unconscious homeostatic mechanisms.

Relevance to Androids

If you're into overclocking central processing units (CPUs), then you're familiar with the menagerie of heat pipes and water-cooled systems available for taming component-killing heat buildup. And water, as well as other liquids, has been used for decades to cool mainframe computers and other heat-generating electronics. Water has a higher heat capacity than air, meaning that you don't need a dozen squirrel-cage blowers to cool a hot-rodded CPU but can get away with a more compact fluid-based system.

Going with the assumption that we are more comfortable dealing with our own kind, then a service android with warm skin and nearly silent operation would be better received than a cold robot with rows of noisy fans for a torso. So, unless there are fundamental advances in the efficiency of electronics, your android is going to produce excess heat. Assuming that the android has an anthropomorphic form, you'll need a way to move heat from the core to the outer surface, mimicking the heat-transfer mechanism of the human body.

Of course, if your goal is simply to give your android faux body heat, then the simplest solution is to embed electric heating pads below the skin or exoskeleton—see the bonus project at the end of this chapter. You'll still need to understand the discussion that follows. Besides, we'll revisit fluids in Chapter 4, where you can apply your experience from the main projects discussed here.

Thermal Model

A simplified thermal model of an android, human, or just about anything with a core temperature can be built with a capacitor, charger, and a few resistors, as in Figure 3-2. In analyzing the model, think of temperature as voltage. Following this

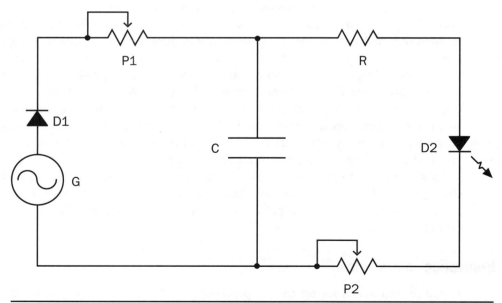

Figure 3-2 Simplified thermal model of an android or human with heat generation (*left*) and loss (*right*).

analogy, the voltage across the capacitor C represents the core body temperature, and the capacitance of C represents the thermal mass of the body.

Capacitor C is discharged by light-emitting diode (LED) D2, which radiates energy in the form of heat and light into the environment. Resistor R represents the thermal resistance from the core to the periphery. The greater R, due to, for example, insulation, the less energy is available for dissipation by the LED. Because of the voltage drop across R—which represents a thermal gradient from core to periphery—the voltage across D2 is less than that across capacitor C. This parallels the lower temperature in the periphery. In humans, this thermal gradient from core to periphery is about 4°C. In an android, the gradient depends on the construction of the android and the design of the heat-exchange system.

Variable resistor R3 represents a means of dynamically adjusting the discharge rate of capacitor C through the periphery, and potentiometer P1 allows adjustment of charge rate. Generator G, diode D1, and P1, the charging circuit, represent body metabolism.

At equilibrium, the charging current from generator G and discharge current through LED D2 are equal. If the load suddenly increases because D2 is shorted (akin to jumping into an ice bath) or because of failure of generator G (skipped breakfast), then the discharge of C can be minimized by increasing P2 (shunting blood from the skin to the organs). Conversely, if C is overcharging, then the excess energy can be sent to D2 for dissipation reducing the value of P2. The challenge is determining when and for how long to increase or decrease P2. This is where control-systems technology comes in.

Simple Negative-Feedback Control

To get a better handle on control systems, let's assume that we have an android with a homeostatic system that consists of a reservoir of liquid, a heat exchanger to dissipate the heat from the liquid, a pump to move liquid from the reservoir to the radiator or heat exchanger and back to the reservoir, and a controller and temperature probe, as in Figure 3-3.

Let's start with simple on-off negative-feedback loop to control the speed of a water pump that directs heated water in the reservoir (core) to the radiator (periphery), where it can dissipate heat to the environment. Let's define the on-off control of the pump with the following pseudocode:

```
If Temperature >= SetPoint, then pump_speed_max.
If Temperature < SetPoint, then pump_off.
```

where `Temperature` is the current temperature of the liquid in the reservoir, and `SetPoint` is the desired temperature of the liquid. Notice that we've defined a typical negative-feedback loop. There's the variable we're trying to control—temperature—and the variable that will be adjusted by the control mechanism—speed of the pump.

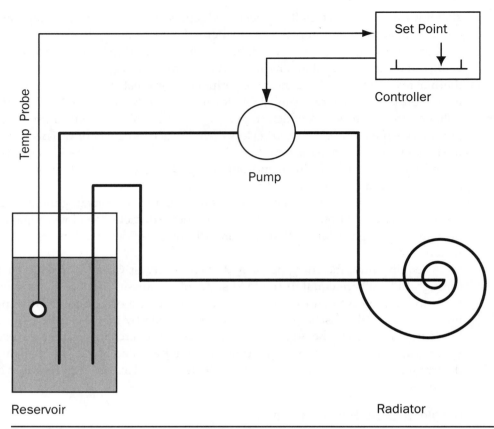

Set Point

Controller

Temp Probe

Pump

Reservoir   Radiator

FIGURE 3-3   Simplified thermal homeostatic system.

Intuitively, an on-off control algorithm isn't going to provide the smoothest approach to the target temperature. An alternative is to use variable control of the pump speed. In pseudocode, the negative-feedback loop equation with variable control of pump speed looks like

```
If Temperature >= SetPoint, then pump_speed (Temperature −
                                        SetPoint).
If Temperature < SetPoint, then pump_off.
```

At issue is how to express the difference between `Temperature` and `SetPoint`. The literal difference in values could be used to specify pump speed directly. Alternatively, the pump speed could be scaled to a percentage of maximum pump speed. Or the difference could be mapped to a nonlinear function that emphasizes the larger difference between `Temperature` and `SetPoint` and that provides only minor corrections in pump speed when the difference is small.

To get a better handle on the feedback loop, we'll convert the pseudocode of on-off control to a more formalized equation that includes real-world variables:

$$Temp_t = Temp_{t-1} + (Temp_{t-1-DT} < SetPoint) \times Inc - R$$

In this equation, DT is dead time, the time required for the pump and water to respond to an "on" signal. Dead time accounts for liquid inertia and the relatively minor delays caused by electrical components. The temperature increment Inc, which represents the incremental change in temperature due to the activity of the pump, is a function of pump design and the pump settings. Resistance R is a catch-all term that includes opposition to liquid flow through the source and return lines and pump, as well as inefficiencies in the pump and radiator.

Note that the second term in the equation is a logical assessment that results in either a true (1) or false (0) value. If the assessment is false, the new temperature is simply the previous temperature value minus the contribution from the resistance of the system.

As noted earlier, a potential problem with a simple on-off control system is overshooting of the set point, as illustrated by the plot in Figure 3-4. Following an initial bolus of hot water, the controller kicks in the pump with a temperature of over 100°F. However, the pump doesn't cut off until the water has been cooled to

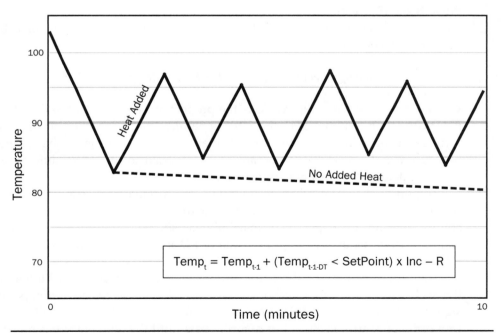

Figure 3-4 Typical on-off negative-feedback control response. Solid line assumes that heat is constantly entering the system; dotted line assumes initial heat only. *SetPoint* = 90°F. Ambient = 70°F.

about 83°F. From a practical perspective, such an overshot might be tolerable for the circuitry housed in an android's torso, but why settle for sloppy control when much better control is only a few hundred keystrokes away?

The tracing in Figure 3-4 assumes that as in the thermal model depicted in Figure 3-2, heat is moving both into and out of the core or thermal reservoir. If there is a finite initial core temperature with no added heat, then after the initial response and the pump stops, the core temperature will slowly decay as a function of the insulation of the reservoir. We show the case of heat constantly entering the system to highlight the nature of the control response. Also, the total time of 10 minutes is arbitrary and depends on the nature of the system, but it allows for a quick comparison of control methodologies.

Proportional-Integral-Derivative Control

An alternative to the simple on-off pump control that has the potential—but no guarantee—to more closely attain set-point temperature than the previously discussed simple control mechanisms is some form of proportional-integral-derivative (PID) controller. The three components of PID control can be combined in various ways and with different emphases on each component. That is, a proportional (P) controller may be sufficient for one application, whereas a proportional-integral (PI) controller is called for in another.

Proportional (P) Control

Taking each component of the PID controller in turn, the equation for a basic proportional control takes the form

$$Temp_t = Temp_{t-1} + (SetPoint + SPC - Temp_{t-1-DT}) \times PC$$

where *Temp* is the temperature, *PC* is the proportion constant, and *SPC* is the set-point correction. Without *SPC*, the *SetPoint* would never be reached. Think of it as a fudge factor to make up for the fact that only a proportion of the difference between actual and target temperature—as opposed to the entire difference—is used to calculate the current temperature.

Proportional control, when used with the appropriate proportion constant and set-point correction, can provide much smoother temperature control than on-off control after an initial period of instability, as shown in Figure 3-5. Tuning a P controller consists of adjusting the value of *PC* and observing the stability of the results.

The typical proportional-control response illustrates a potential limitation of the technique—the initial response may overshoot the set-point temperature significantly. In this example, immediately following a thermal challenge, the proportional control kicks in but doesn't abate until the temperature is 6 degrees lower than the set point. Note that the time axis is arbitrary and depends on ambient conditions as well as the controller constants.

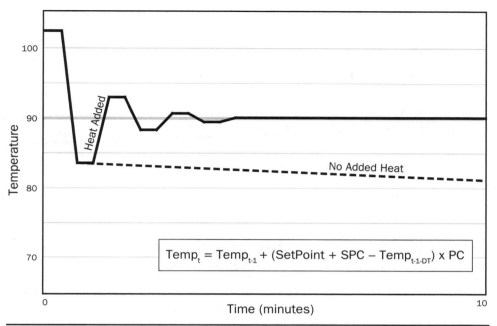

Figure 3-5 Typical proportional-control response. Solid line = heat is constantly entering the system; dotted line = initial heat only. *SetPoint* = 90°F.

Proportional control is characterized by instability at specific combinations of set point, set-point correction, dead time, and proportion constant values. In particular, as the proportion constant is increased, there is a tendency for the temperature to oscillate.

Proportional-Integral (PI) Control

Proportional-integral control addresses the instability of the purely proportional-control scheme by adding a term that represents the sum of the difference between desired and actual temperature over time. The basic proportional-integral control equation is defined as

$$Temp_t = Temp_{t-1} + (SetPoint + Temp_{t-1-DT}) \times$$
$$PC + Integral_{t-1} + (Temp_t - Integral_{t-1}) \times IC$$

where *Integral* is the sum of differences between *SetPoint* and current temperature, and *IC*, the integral constant, determines the contribution of the integral to the current temperature. Note the absence of *SPC*, the set-point correction, in the equation.

Compare the response curve for PI control, shown in Figure 3-6, with that of the curve for P control in Figure 3-5. Even though the Microsoft Excel models used to

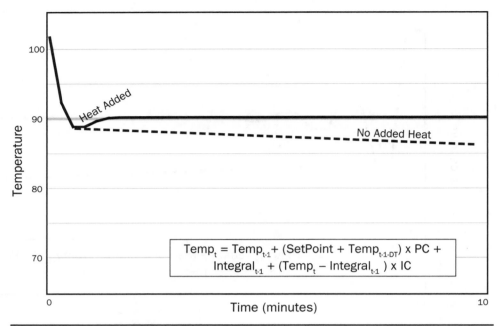

$$Temp_t = Temp_{t-1} + (SetPoint + Temp_{t-1-DT}) \times PC + Integral_{t-1} + (Temp_t - Integral_{t-1}) \times IC$$

FIGURE 3-6   Typical PI controller response. Solid line assumes that heat is constantly entering the system; dotted line assumes initial heat only. *SetPoint* = 90°F. Ambient = 70°F.

generate the two graphs have identical values for set point, set-point correction, dead time, and proportion constant, PI control results in a smaller overshoot and relatively quicker stabilization. Note that the 10-minute timeline in Figure 3-6 is approximate but comparable with the other plots of temperature versus time in this chapter.

Because the integral term reflects the difference between actual and desired temperatures over time, proportional-integral control is most useful when the ambient temperature is relatively constant. Conversely, the addition of the integral term is less beneficial when the environmental conditions are constantly shifting. Tuning a PI controller, which consists of identifying the optimal values for the proportion constant *PC* and the integral constant *IC*, is often more of a challenge than tuning a proportional controller.

Proportional-Integral-Derivative (PID) Control
Like the purely proportional controller, the proportional-integral controller is predisposed to instability with some variable combinations. To counteract this instability, a derivative term—the equivalent of a smoothing capacitor in a power supply—can be used to suppress the rate of change. The full proportional-integral-derivative control equation is defined as:

$$Temp_t = Temp_{t-1} + (SetPoint + Temp_{t-1-DT}) \times PC + Integral_{t-1}$$
$$+ (Temp_t - Integral_{t-1}) \times IC - (Temp_{t-1} - Temp_t) \times DC$$

where *DC*, the derivative constant, determines the sensitivity of the system to second-to-second variations in temperature. Unlike the proportional and integral terms, the derivative term is subtracted from previous temperature value. This tends to diminish the rate of change, as shown in Figure 3-7. As you can see, unlike the previous controllers, there is no overshoot in the response curve.

The four previous graphs, while typical, are idealized, in that the constants are selected to illustrate optimal responses. That is, the controllers are tuned. Tuning a PID controller consists of optimizing the values of the three constants, proportion constant *PC*, integral constant *IC*, and derivative constant *DC*, which can be a formidable task with a complex system. Fortunately, our system is relatively simple.

Even though PID seems superior to the simple controller with variable response, implementing a PID controller doesn't guarantee a rapid move to the set point with no overshoot. To obtain results that mirror Figure 3-7, you have to be smart about adjusting the constants. Another requirement is that the system under control is repeatable; otherwise, the constants have to be continually redefined. You can help to ensure repeatability by using the same initial conditions, such as reservoir

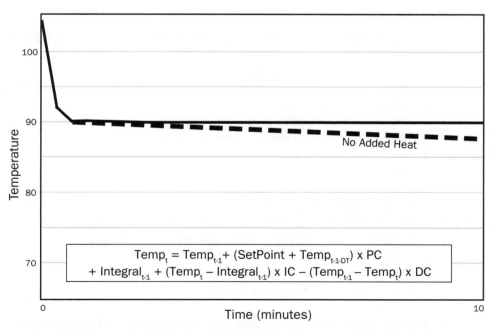

FIGURE 3-7   Solid line assumes that heat is constantly entering the system; dotted line assumes initial heat only. *SetPoint* = 90°F. Ambient = 70°F.

temperature, reservoir volume, ambient temperature, humidity, and airflow around the reservoir and coils.

In addition to these basic equations, there are numerous variations and enhancements to PID controllers, as well as more sophisticated algorithms for motion control. Although it's critical to understand the advantages and limitations of the PID components discussed earlier, we can avoid the details of implementation by using a standard PID library for the Arduino.

Thermal Homeostasis Simulator

The goal in the following set of experiments is to provide you with hands-on experience with the precding control models in the context of thermal homeostasis. The principles you learn here should be applicable to a variety of control systems, whether the target is an android or some other mechatronic device.

Basic Homeostasis Simulator

We start with a simple thermal heat-exchange system similar to the one illustrated earlier in Figure 3-3. Assuming that you don't have an android handy with a torso capable of holding a reservoir of hot liquid, we'll use a disembodied approach with separate components. This has the advantage of supporting quick modifications to the setup and easy visualization of the homeostatic process.

Bill of Materials

To construct the basic thermal homeostatic simulator, you'll need the following:

- 6- to 12-V dirent-current (dc) water pump
- 10 feet of 3/16-inch silicone aquarium airline or 1/8-inch copper tubing
- Tube of silicone sealant
- Hot water (from tap) or immersion heater
- Fittings or sealant to attach airline to the pump
- Insulated 1-gal container
- Arduino Uno or equivalent
- USB connection for the Arduino
- Power supply for the pump
- Motor-control shield or N-channel MOSFET rate at 2 A
- 10-kΩ thermistor in waterproof tube or container
- 10-kΩ, ¼-W resistor
- Mannequin head or other platform
- Can of compressed air or heat sink

As with previous projects, feel free to substitute where economics and availability dictate. For example, an immersion heater is a convenient way to heat the water in the reservoir, but hot water from the tap works just as well. Also, there's nothing special about the gallon container. For our purposes, a stainless steel or aluminum pot wrapped with a towel or bubble wrap works just as well as a supersize laboratory beaker with a fiberglass jacket. Once you've debugged the system, a sealed paint can makes an inexpensive deployment container.

The critical items in the preceding list are the low-voltage dc pump, airline, and associated fittings. The emphasis here is low-voltage dc for both safety and ease of control. Check out the affordable dc water pumps, both submersible and with input-output hose connections sold for bilge pumps and for small water fountains. As shown in Figure 3-8, we use a submersible 12-Vdc @ 1.2-A bilge pump rated at 350 gal/h available for about $16. The 350 gal/h output of this pump is a bit of overkill for this application, but you'll want something rated at moving at least a liter or two of water per minute through 10 feet of 3/16-inch hose.

Also shown in Figure 3-8 are 3/16-inch silicone aquarium airline, a MOSFET shield with added screw shield for easy prototyping, a universal tubing adapter, and two thermistor probes—one do-it-yourself (DIY) and one commercial. The copper tubing may be substituted for the airline, and the IR thermometer, while not

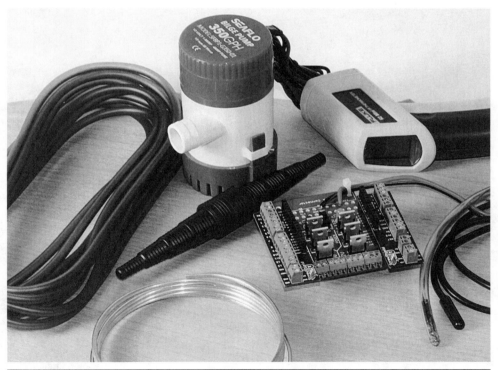

FIGURE 3-8 Major components used to construct and operate the homeostatic system. Copper tubing and infrared (IR) thermometer are optional.

necessary, is nice to have. It will enable you to sweep the entire system in seconds to check component temperature without the need for direct contact.

We'll use the silicone airline to construct the radiator. Granted, silicone is not the best choice for a heat exchanger, but it's cheap and easier to work with than copper tubing. Unless you know what you're looking for, you could easily spend an afternoon in a hardware store searching for the correct brass fitting. However, feel free to substitute copper or aluminum tubing for silicone if you have it and know how to use it. It will provide much faster response times.

Because of the Arduino's 40-mA limit per port, we need a high-power transistor or H-bridge to power the pump. Figure 3-9 shows some options based on both MOSFETs and an integrated bridge chip. The Sparkfun motor shield, on the left, uses six MOSFETS to provide six high-current channels. The Grove shield, on the right, uses an L298N dual full-bridge driver. A single N-channel MOSFET with internal diode protection, such as the Fairchild RFP30N06LE, together with heat sink, PC board, and input-output connectors, shown in the center of the figure, is the most compact and affordable solution for a single dc motor. If you decide to use a single MOSFET, use a PC board and heavy-duty connectors. If you try to use a breadboard, you'll risk burning out the tracks.

Figure 3-9 Motor-control hardware options include a MOSFET-based shield (*left*), a single MOSFET (*center*), and a bridge driver-based shield (*right*).

Circuit

The circuit, shown in Figure 3-10, consists of an Arduino with a 12-Vdc pump driven by an *N*-channel MOSFET, and the thermistor probe feeding an analog input. The thermistor and 10-kΩ resistor R1 form a voltage divider that provides a greater input to port A0 with increasing temperature. Recall that thermistors are passive and have a negative temperature coefficient, so resistance decreases with increasing temperature. As a result, with increasing temperature, the relative voltage drop across R1 increases, resulting in a higher reading on the Arduino.

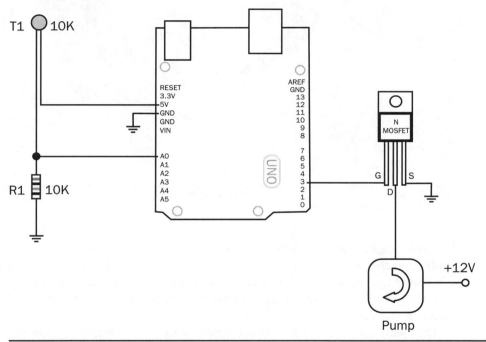

FIGURE 3-10   Thermal homeostatic circuit.

Construction

As you can see from the schematic, there isn't much to do in terms of electronics construction. If you opt to use a 10-kΩ thermistor for the temperature probe, then you'll need to make a waterproof housing. Insert the thermistor into a 1-foot section of airline or vinyl tube, followed by silicone sealant to keep the electrical connection dry. Use ample shrink-wrap tubing on the leads and solder connections to prevent shorting and to form a barrier against water and moisture. Figure 3-11 shows the main materials required to create a waterproof thermistor probe.

A rubber or plastic cap will make a better seal, at the expense of response time, and crisp temperature-sensor response time is critical for accurate control of the system. The temperature of a waterproof, nonconductive bulb covering the

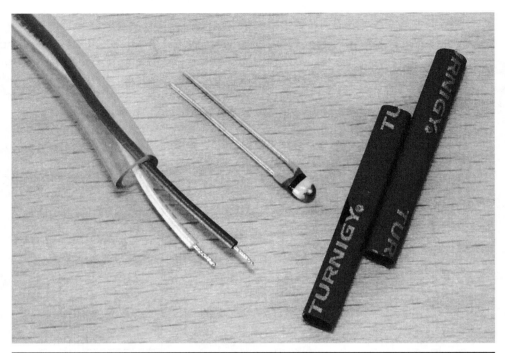

Figure 3-11 Components used to construct a waterproof thermistor probe: twisted pair, flexible vinyl tube, and shrink-wrap tubing.

thermistor lags behind the water as a function of the covering's mass and thermal-conduction characteristics. If you're not into DIY probes, you can buy prefabricated waterproof thermistor-based probes that have thin, thermally conductive caps and good response times. A completed DIY probe and commercial thermistor probe are shown in Figure 3-8.

The real construction challenge deals with the plumbing, including the supporting platform. Ideally, the tubing would be run underneath a thin membrane of vinyl skin, mimicking prominent veins. For illustration purposes, though, we use an inexpensive mannequin head and wrap the silicone tubing around the head like a sweatband, as in Figure 3-12. The tubing can be covered with a light wig for a more natural look.

An alternative is to use copper tubing—a much better conductor and radiator of heat—on a mannequin head or other platform. For example, Figure 3-13 shows copper tubing mounted on an exercise roller. Note that the connection to the pump is flexible airline. Regardless of the type of tubing or support platform, you'll probably find that the most difficult part of construction is connecting the tubing to the water pump. For example, you likely will need a tubing size converter, such as the one shown at the start of this chapter, if the pump outlet outside diameter (OD) and tubing inside diameter (ID) don't match.

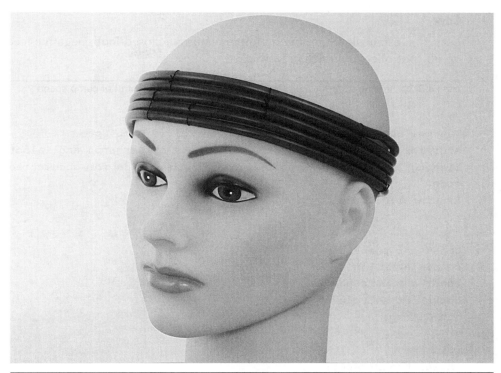

FIGURE 3-12   Silicone tube radiator mounted on mannequin head.

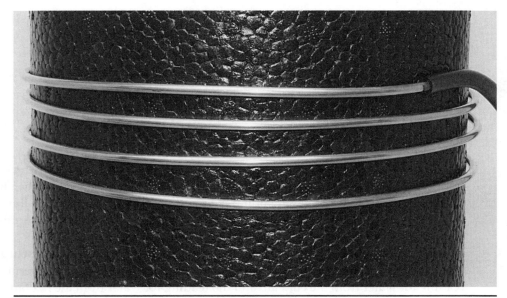

FIGURE 3-13   Copper-tube radiator mounted on foam exercise roller.

Code

Let's first establish a baseline by implementing a closed-loop, negative-feedback on-off controller, as shown in Listing 3-1.

LISTING 3-1 Arduino code for thermal hemostasis using on-off control of pump speed.

```
/*
Thermal Homeostasis Using On-Off Control of Pump Speed
Androids: Build Your Own Lifelike Robots by Bergeron and Talbot
Listing 3-1. See www.mhprofessional.com/Androids for documented
code
Arduino 1.0.1 environment
*/

const int pumpPin = 3;
const int T1Pin = A0;
const int pumpSpeed = 100;
const int setPoint = 600;
int T1Temp;

void setup() {
Serial.begin(9600);
}

void loop() {
  T1Temp = analogRead(T1Pin);
  Serial.println(T1Temp);
  if (T1Temp > setPoint)
  analogWrite(pumpPin,pumpSpeed);
else
  analogWrite(pumpPin,0);
  delay(1000);
}
```

This example relies on data sent to the serial monitor by the `Serial.println()` function. If the reading from the thermistor indicates a temperature higher than `setPoint` temperature, then the pump is turned on, and water circulates through the radiator, where water gives up heat to the environment. When the thermistor indicates a temperature of `setPoint` or less, then the pump stops.

The optimal values for `setPoint` and `pumpSpeed` depend on your setup. In our experiments, a `setPoint` of 600 is about midway between the thermistor reading for water from the cold tap and the reading for water from the hot tap. The optimal value for `pumpSpeed` depends on the design of your system as well as the response you're looking for. If you're using a high-throughput pump, it's safer to err on the side of a smaller `pumpSpeed`.

Assuming that your connections can handle the stress, with the pump operating at full speed, core temperature should drop rapidly. However, it's also more likely to overshoot the `setPoint` temperature. On the other hand, if you slow the pump to a trickle, then it will take longer to dissipate the heat from the reservoir but is less likely to overshoot the `setPoint` temperature. Obviously, a bit of trial and error is called for.

Operation

To get things moving, either fill the reservoir with hot tap water or power up the immersion heater. In either case, the water in the reservoir has to be at a temperature greater than the set point. In addition, you should specify a `setPoint` value that corresponds to a reasonable temperature—you don't want to melt the plastic parts of your pump, for example, and yet you want the temperature of the water to be significantly greater than ambient. Otherwise, your radiator isn't going to do much radiating.

If you're using an immersion heater instead of hot tap water, then an IR or direct-contact thermometer is helpful to determine when to unplug the heater. You don't want to boil the water pump. Other than guarding against excessive temperatures, measuring the absolute temperature isn't necessary. The goal is to highlight patterns in the thermistor readings so that you can compare them with those resulting from more advanced control algorithms.

If you're having trouble with this or any of the following systems, you'll save time by debugging in air instead of water. Remove the thermistor probe from the reservoir, and dry it. Next, change `setPoint` so that it corresponds to a temperature just above ambient temperature. Squeeze the probe in your hands to raise the temperature of the probe above the value of `setPoint`, and use the compressed air to lower the temperature. Check the connections to the pump as well as the integrity of the thermistor probe. If you didn't manage to seal the thermistor, then water on the leads could be responsible for abnormal behavior.

Once you've verified that the probe and associated circuitry are working properly, return `setPoint` to its original value. If you're still having problems, then check the pump and associated circuitry. Note that if you're debugging the water pump, some submersible units require continuous contact with water for lubrication and heat removal.

Modifications

Graduated Pump Speed Control

We should be able to avoid significantly overshooting the value of `setPoint` by varying pump speed as a function of the difference between actual and target temperatures.

Code

The Arduino code for the system, shown in Listing 3-2, uses the standard map() scaling function to establish the value assigned to pumpSpeed. You'll need to modify the value of the variables used in the map function to fit your system. For example, in our case, the temperature of water from the hot tap, hotTemp, corresponds to a thermistor reading of 760. The value you select will depend on your hot-water heat settings and the response of your particular thermistor. Similarly, pay attention to pumpSpeedMax, the maximum pump speed. Start out with a small value if you have any doubt about the integrity of your plumbing job.

LISTING 3-2   Homeostatic control using a graduated response.

```
/*
Thermal Homeostasis Using Variable Control of Pump Speed
Androids: Build Your Own Lifelike Robots by Bergeron and Talbot
Listing 3-2. See www.mhprofessional.com/Androids for documented
code
Arduino 1.0.1 environment
*/
const int pumpPin = 3;
const int T1Pin = A0;
const int pumpSpeedMax = 100;
const int setPoint = 600;
const int ambientTemp = 512;
const int hotTemp = 760;
int T1Temp;
int pumpSpeed;

void setup(){
Serial.begin(9600);
}

void loop() {
  T1Temp = analogRead(T1Pin);
  Serial.println(T1Temp);
  pumpSpeed = map(T1Temp, setPoint, hotTemp, 0, pumpSpeedMax);
  if (T1Temp > setPoint) {
    Serial.println(pumpSpeed);
    analogWrite(pumpPin,pumpSpeed);
  }
      else
  analogWrite(pumpPin,0);
  delay(1000);
}
```

Note that the variability in the graduated response is limited to the times when the thermistor response to the water temperature, `T1Temp`, is greater than `setPoint`. Otherwise, the `pumpSpeed` is set to zero.

Operation

Setup and operation are the same as with the on-off control experiment. Either add a fixed amount of hot water to the reservoir, or use the immersion heater to raise the temperature in the reservoir above the temperature corresponding to the value of `setPoint`. Ideally, the overshoot with graduated pump speed control should be less pronounced than with on-off control.

PID Control

You should find the regulation provided by graduated speed control superior to on-off control. The question is, is it good enough, and if not, can we improve on it? For this application, simple graduated control may be sufficient— and that's worth keeping in mind when you're faced with the added complexity and resource requirements associated with PID control.

Given that we're not using the Arduino to monitor a dozen other sensors, there's random access memory (RAM) and processing power to spare, but this may not always be the case. For now, let's assume that simple graduated control isn't good enough and that we have ample computational resources.

Code

The Arduino code for the PID system, shown in Listing 3-3, is only modestly more complex than Listing 3-2, thanks to the PID library from the Arduino Playground. This library hides many of the implementation details we discussed earlier, allowing you to focus on the optimal values for point correction *PC*, integral constant *IC*, and derivative constant *DC*, as well as operation of the overall system.

LISTING 3-3   Thermal homeostasis with a PID-based controller.

```
/*
Thermal Homeostasis Using PID Controller
Androids: Build Your Own Lifelike Robots by Bergeron and Talbot
Listing 3-3. See www.mhprofessional.com/Androids for fully
documented code
Arduino 1.0.1 environment
*/

#include <PID_v1.h>
const int pumpPin = 3;
const int PC = 2;
const int IC = 5;
```

```
const int DC = 1;

double Setpoint = 470;
double MaxOutput = 100;
double Input, Output;

PID myPID(&Input, &Output, &Setpoint, PC, IC, DC, REVERSE);

void setup(){
  Serial.begin(9600);
  Input = analogRead(0);
  myPID.SetMode(AUTOMATIC);myPID.SetOutputLimits(0,MaxOutput);
}

void loop(){
  Input = analogRead(0);
  Serial.println(Input);
  myPID.Compute();
  Serial.println(Output);
  analogWrite(pumpPin,Output);
}
```

The cornerstone of this program is the `myPID()` function that defines the `Output` (motor speed) given the `Input` (thermistor reading). Tuning the PID algorithm entails substituting new values for the three constants *PC*, *IC*, and *DC*. The values provided in the listing are a good place to start.

Operation

The system operates the same as in the previous two examples, but let's get more specific in establishing initial conditions. Instead of simply adding hot water or plugging in the immersion heater until `setPoint` is exceeded, either add a fixed amount of hot water at a known temperature to the reservoir, or plug in the immersion heater for a fixed number of seconds. Stir the water in the reservoir, and then track the thermistor reading, `Input`, and the drive level, `Output`, to the water pump. For debugging purposes, you can temporarily add a delay statement in the main loop so that you can read the values listed for `Input` and `Output` in real time. The delay will alter the response of the PID controller.

If your results aren't better than with simple control—meaning rapid movement to the set-point temperature with little or no overshoot—then you may need to tune the PID. PID tuning utilities for Arduino are available for the task, and this system is simple enough that trial and error are a viable option. A popular PID tuning utility for the Arduino is the appropriately named PID autotune library, available at the Arduino Playground.

Another possibility is that your system is too unstable for a PID controller. There may be differences in water temperature and volume with iterations of the experiment, inadequate insulation of the reservoir, changes in ambient temperature, ineffective heat transfer from the radiator, or random leaks in the plumbing. Whatever the reason, if the basic system doesn't support repeatable results, then you'll never get the constants dialed in.

Also, recall from the discussion of the derivative component of the PID that the entire derivative term is a function of the constant *DC*. If you set *DC* to zero in Listing 3-3, you'll have a PI controller. Even if the full PID controller is working for you, it's a good idea to work with the PI controller to get a feel for tuning the system with only the *PC* and *IC* constants.

Peripheral Temperature Response

Let's add some biological mimicry to the system by making the core-temperature `setPoint` responsive to the temperature at the periphery. For this modification, we'll add a second thermistor probe and use the readings to modify `setPoint`.

Bill of Materials

In addition to the basic PID system, you'll need the following:

- 10-kΩ thermistor
- 10-kΩ, ¼-W resistor
- Desk lamp, blow dryer, or other heat source
- Can of compressed air or heat sink

Circuit

The circuit, shown in Figure 3-14, consists of the original circuitry plus a second thermistor-resistor pair T2-R2 to monitor the temperature of the mannequin head or other support. The 10-kΩ resistor R2 is used to create a voltage divider that results in higher readings at the Arduino with increasing temperature.

Construction

Connect a twisted pair of wire to the second thermistor, and tape or otherwise secure the thermistor to the mannequin's neck or other support. Attach the thermistor so that the component can be easily accessed to apply heat or cold.

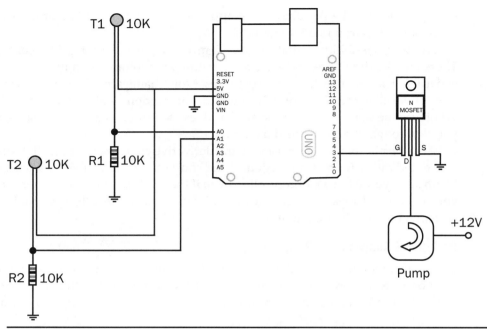

Figure 3-14   Circuit with core (T1-R1) and peripheral (T2-R2) temperature probes.

Code

The Arduino code for the dual-thermistor circuit is shown in Listing 3-4. Again, because we leverage the Arduino PID library, there isn't much to the program. If the reading from T2 is elevated, then `setPoint` is depressed, meaning that the pump will kick in sooner, starting the cooling process earlier. Conversely, if you spray compressed air on T2, lowering its temperature, then `setPoint` will be elevated. The pump won't start until T1 produces a relatively higher reading.

Listing 3-4   Arduino code for a dual-thermistor homeostatic system.

```
/*
Thermal Homeostasis Using PID Controller
Androids: Build Your Own Lifelike Robots by Bergeron and Talbot
Listing 3-4. See www.mhprofessional.com/Androids for fully
documented code
Arduino 1.0.1 environment
*/

#include <PID_v1.h>
const int pumpPin = 3;
const int T1Pin = A0;
```

```
const int T2Pin = A1;
const int PC = 2;
const int IC = 5;
const int DC = 1;

double SetPoint = 470;
double MaxOutput = 100;
double SetPointMid = 470;
double SetPointHigh = 490;
double SetPointLow = 460;
double Input, Output;
int ambientTemp;
int triggerTemp = 10;

PID myPID(&Input, &Output, &SetPoint,PC, IC, DC, REVERSE);

void setup(){
  Serial.begin(9600);
  Input = analogRead(T1Pin);
  myPID.SetMode(AUTOMATIC);
myPID.SetOutputLimits(0,MaxOutput);
}

void loop(){
  Input = analogRead(T1Pin);
  ambientTemp = analogRead(T2Pin);
  Serial.println(Input);
  Serial.println(ambientTemp);
  if ((ambientTemp - Input)>triggerTemp){
  SetPoint = SetPointHigh;
}
else if ((Input - ambientTemp)>triggerTemp){
  SetPoint = SetPointLow;
}
else{
  SetPoint = SetPointMid;
}
  myPID.Compute();
  Serial.println(Output);
  Serial.println(SetPoint);
  analogWrite(pumpPin,Output);
}
```

The major change in this listing, compared with the preceding listing for a single-thermistor probe, is the setup and execution of assessing the difference between the values returned by the two thermistors. Note that in this example the spread between `SetPointHigh`, `SetPointMid`, and `SetPointLow` is asymmetrical. You can and should experiment with symmetrical spreads, as well as wider and narrower spreads between the values assigned to these variables.

Operation

By now, you should know the drill. Carefully add hot water or plug in the immersion heater in a manner such that you can replicate the initial conditions. Then run the experiment and track the data printout. Using a dynamic value for `setPoint`, while mimicking the physiological response in humans, also perturbs the PID controller. If you change the `setPoint` value significantly, you may need to also dynamically assign new constants for the PID algorithm using the `SetTunings()` function in the PID library.

Bonus Project: PID Control of Electric Heating Pad

Okay, let's say that you're not into mixing water and electricity. You can still apply the preceding theory and code to an all-electric thermal system with only minor modification of the code. If your goal is to simply provide body heat, without regard for removing heat generated by your android, then the easiest approach is to combine the PID control code in Listing 3-3 with heating elements below the skin of your android.

Bill of Materials

For the bonus project, you'll need the following:

- Arduino Uno or equivalent
- 5-Vdc polyimide-film heating pad(s)
- 9- to 12-Vdc power supply for the pad(s)
- Motor-control shield or N-channel MOSFET rated at 2 A
- 10-kΩ thermistor
- 10-kΩ, ¼-W resistor
- Polyvinyl chloride (PVC) tube or other platform

In short, we've done away with the tubing, pump, and other water-related parts and have substituted a fancy resistor, namely, the flexible, lightweight, and inexpensive 5- × 15-cm polyimide-film heating pads from Sparkfun shown in Figure 3-15. The paper-thin pads, which incorporate a mesh of polyester filament

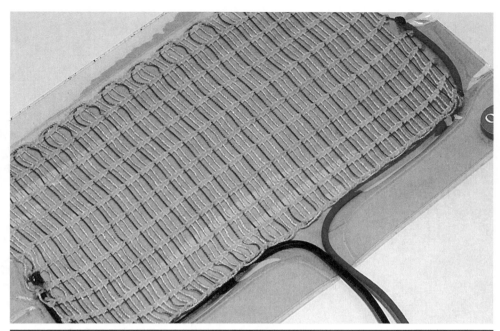

Figure 3-15   Close-up of a polyimide-film heating pad.

and metal conductive fiber, draw about 600 mA at 5 Vdc and get up to about 120°F after a few minutes. If surface area is an issue—such as heating the surface of an android hand for that warm handshake—then consider the 5- × 10-cm heating pads. The smaller pads draw about 850 mA at 5 Vdc.

Circuit

The circuit, shown in Figure 3-16, consists of an Arduino with a heating pad driven by an *N*-channel MOSFET and the thermistor probe feeding an analog input. If you've built the basic pump circuit, simply replace the pump connections with those for the heating pad. A 5-Vdc supply will work, but a 9- or 12-Vdc supply will provide a greater range of temperature control and won't harm the heating pad.

The pads are simply resistors. They can be wired in series or parallel depending on your power supply, motor-shield capabilities, and number of pads required for your application. The downside of wiring everything in series or parallel is that you forfeit the ability to control the temperature of individual pads.

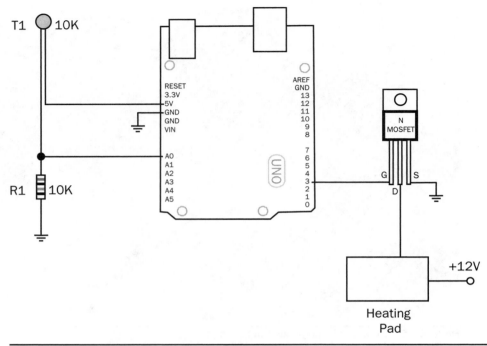

FIGURE 3-16   Thermal homeostatic circuit using a heating pad.

Construction

Simply tape the heating pad where it's convenient—to a PVC pipe the diameter of your arm or even to your desktop. The only real constraint is the placement of the thermistor. It should be on or near the surface of the pad. For example, if you tape the pad to a 3-inch-diameter pipe and then cover the pipe with vinyl—to approximate an arm—then drill a hold for the thermistor, and mount it on the underside of the heating pad.

Code

The Arduino code for this experiment is identical to that in Listing 3-3. As with the previous projects, you'll have to experiment with values for point correction *PC*, integral constant *IC*, and derivative constant *DC* to fit the characteristics of your particular setup.

Operation

This is a plug-and-play system. What you feel—in terms of temperature—is what you get. Because these heating pads are designed for use as body warmers for winter clothes, there isn't much chance that you'll toast your fingers. But use an IR thermometer if you have one to monitor the actual temperature, just in case.

Gremlins

As you work through the projects, remember that even though you're dealing with relatively low voltage, you can still get a nasty shock. Keep your hands dry and away from the reservoir. It's also a good idea to assume that the fittings or the airline will burst at some point, creating a miniature Water Wiggle on your desktop. You should have a safe means of quickly disconnecting the power to the system and a stack of towels handy to clean up after any spills. And if you're not confident in your plumbing skills, consider running the experiment in or near your kitchen sink.

If you decide to deploy one of the water-based systems on a robotic platform, then you'll probably want to create a closed system. As noted earlier, an empty 1-gal paint can and lid can be used to make an inexpensive sealed container. A sealed system is more involved than the open system discussed earlier, in that you'll have to mount through-wall connectors and install a pressure-relief valve and a water-level detector to complete the system. You should be able to find the necessary plumbing supplies at your favorite local or online hardware store. Aquarium supply stores and sites are also worth visiting.

Search Terms

Relevant search terms for your Web browser include

- PID controller
- Temperature regulation
- Hypothermia
- Hyperthermia

CHAPTER 4

Light and Vision

The ability to see one's environment is critical for survival and interaction. Computer vision for recognizing objects from a camera feed is an advanced science that requires sophisticated algorithms and is usually beyond the scope of what microcontrollers can do. Although it is not possible for an Arduino to recognize people and see them smile, there are many compelling vision applications that can take advantage of simple sensors so that your android can see and adapt to the environment around it.

In this chapter we will explore light sensors that will give your Android some degree of vision. We will also create a pair of eyes with variable pupils and gaze that follows light sources. This will give your android the appearance of a living face for humans to interact with. Necessary parts include an Arduino, a couple of 8 × 8 light-emitting diode (LED) matrix displays, some photoresistors, and a special tracking camera that works with microcontrollers. We will also begin construction of components for an android head that will be completed in Chapter 9.

Biological Basis and Pupillary Light Reflex

The human eye is a fluid-filled globe lined by a photoreceptor, the retina, at the back. The light aperture contains the transparent lens and cornea. The amount of light permitted to enter is controlled by the iris, which can contract and expand to adapt to ambient lighting conditions. The iris is the colored part of the eye, and the opening that allows you to see the retina directly is called the pupil. A person in a bright room or with a flashlight shining in his eyes will constrict his pupils to decrease the amount of light arriving at the retina. The medical term for this whole process is called the *pupillary light reflex*. Of course, sitting in a dark space will eventually cause the pupils to dilate as well.

Light is not the only thing that affects the pupils. In fact, an astute observer can tell quite a bit about someone by observing the amount of pupil dilation. Humans

97

dilate their pupils in response to flight-or-flight situations, sexual arousal, and certain medications or drugs. Pupils also constrict when a person is falling asleep, under the influence of medications, and from some intoxicants. A physician encountering multiple people with very small pupils would be very suspicious for nerve-agent poisoning, for example.

The pupillary light reflex examination provides a lot of diagnostic information to doctors because it allows for testing of sensory, motor, and brain functions related to the eyes. This reflex tells us that all connections related to it within the brain stem are functioning and active. When the examination is performed, the doctor shines a flashlight into one eye. The expected result is that both eyes will constrict quickly and symmetrically. When the light is removed, the pupils will slowly go back to their previous slightly more dilated state. The doctor will then shine a light in the other eye and expect the same results. If the directly examined eye constricts but not the opposite, "consensual" eye or vice versa, then something is seriously wrong with the patient. Think of this examination as sort of a computer boot-code message for humans (Figure 4-1). Humans who have died will have fixed and dilated pupils that lack reflexes. That's the human version of the blue crash screen.

When examining our android, the doctor should be able to shine a light in either eye of the android and get nice, quick pupil constrictions with a slow return to normal after removing the light. The android also will respond to ambient lighting with the right size pupil at all times.

Importance of Eyes for Interhuman and Android-Human Communication

A great deal of human communication is nonverbal. The ability for an android to direct vision, show pupillary responses, and create expressions makes the face real to people and is critical to building robots that people can relate to. The eyes we will use for this chapter's experiment will be square 8×8 LED matrix displays that look quite robotic but get the point across. People tend to find these digital eyes to be "cute" and are usually affectionately inclined toward androids that employ this approach. These eyes will be repurposed and the topic will be expanded on in Chapter 9.

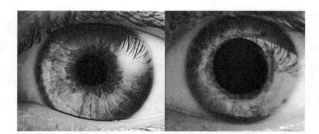

Figure 4-1 Pupillary reflex. Normal versus dilated pupils are demonstrated.

Simple Light-Reflex Experiment

Our first step in building an expressive android is to make the pupil of the eye responsive to the level of ambient illumination. The simple version of this will require a cadmium-sulfide (CdS) photocell, one resistor, and an analog input on your microcontroller. The photocell resistance decreases based on the amount of light. The photocell is arranged with a resistor so as to form a voltage divider. The result of this is that more light reaching the photocell results in a higher voltage detected by the microcontroller. In this case, the Arduino microcontroller will detect some number between 0 and 1,023 that will vary based on the individual properties of the photocell, so you may need to make adjustments to the variables `maxLightRight`, `maxLightLeft`, `minLightRight`, and `minLightLeft`. These variables define the minimum and maximum voltages you are getting from your circuit. We will use these values for the later experiments, so be sure to record the values.

When you have assembled the circuit (Figures 4-2 and 4-3) and then run the code, you can determine the `minLight` values by covering the photocells with your hand or preferably turning out all the lights in a closed room. The `maxLight` values are determined by shining a bright light directly into the photocells and recording the numbers again. It might also be useful to get an idea what the value is for typical ambient light conditions—this will probably be about three-quarters of `maxLight`. The numbers we are using are raw values from the analog-to-digital (A/D) converter circuit on the Arduino rather than literal voltages. If the numbers

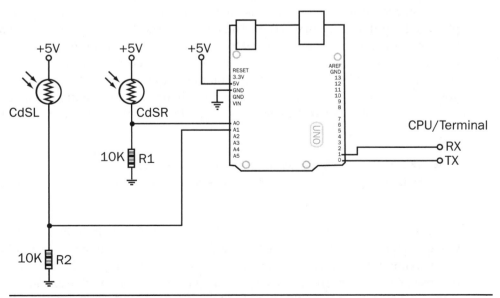

FIGURE 4-2   Schematic of simple light reflex experiment.

FIGURE 4-3   Photograph of simple light reflex experiment.

coming out of this circuit go out of range (freeze at 1,023) or hover near zero, then it may be necessary to adjust the resistor value. The schematic default is 10 kΩ, but 5 or 15 kΩ may work better depending on the resistance range of your photoresistor.

Another feature of this code is that it automatically normalizes your CdS photocell input and calculates the `minLight` and `maxLight` values based on the range of light the sensors are exposed to for each run. When running the calibration utility, do not forget about running it for a few seconds in completely dark conditions as well as with a flashlight pointing into the photoresistors. By placing your hand over the sensors, you can see how changes in illumination change each sensor's raw value (`RT` or `LT`) and that a number is computed that represents an adjusted value for light:

```
rangeOfLight = maxLightRight - minLightRight;
adjustedLightRight = (sensorValue - minLightRight) / (rangeOfLight
                / 100);
```

The resistance range for any CdS photocell is notoriously individualistic, so we will use a bit of math to compute the rough percentage of light between minimum and maximum values. This is computed by calculating `rangeOfLight` (the range of sensed light) and putting it in the preceding formula. The variable `sensorValue` is the analog voltage from the circuit with a range of `0-1023`. Doing this for both

CdS photocells will put the results within 5 to 10 percent of each other if they are both exposed to the same light. In the output example here, there is a bit more light hitting the right eye receptor (left) than the left one. Notice the difference in min/max range between two identical-appearing CdS photoresistors from the same package. In all examples, *left* or *right* refers to the left or right eye as if you were looking at a face, so remember to reverse your point of reference. Running the CdS photocell calibration program will produce an output similar to this:

```
CdS Photoresistor Calibration Program
Androids: Build Your Own Lifelike Robots by Bergeron and Talbot
MIN   RT_   MAX    % % %        MIN   LT   MAX    % % %
140   741   994    75           538   850  1000   78
```

The "% % %" column is the adjustedLight variable. Notice that the right eye CdS photocell has a much lower MIN value than the left eye photocell. Despite the fact that the MIN values are 400 points apart and live readings are more than 100 points apart when exposed to the same light, the normalized value, adjustedLight, is only 3 points different between the two cells. How's that for fancy vision processing?

Code

Listing 4-1   CdS photoresistor calibration utility.

```
/*
CdS Photoresistor Calibration Utility
Androids: Build Your Own Lifelike Robots by Bergeron and Talbot
Listing 4-1. See www.mhprofessional.com/Androids for documented
code.
By Thomas Talbot
Arduino 1.0.1 environment
*/
int sensorRight = A0;
int sensorLeft = A1;
int sensorValue = 0;
int maxLightRight = 0;
int minLightRight = 1023;
int adjustedLightRight = 0;
int maxLightLeft = 0;
int minLightLeft = 1023;
int adjustedLightLeft = 0;
int rangeOfLight = 0;
```

```
void setup() {
  Serial.begin(9600);
  Serial.println("CdS Photoresistor Calibration Program");
  Serial.println("Androids: Build Your Own Lifelike Robots by
                  Bergeron and Talbot");
}

void loop() {
  sensorValue = analogRead(sensorRight);
  if (maxLightRight < sensorValue) maxLightRight = sensorValue
  if (minLightRight > sensorValue) minLightRight = sensorValue;
  rangeOfLight = maxLightRight - minLightRight;
  adjustedLightRight = (sensorValue - minLightRight) /
                          (rangeOfLight / 100);
  Serial.println("-");
  Serial.println("MIN  RT_  MAX    percent percent percent
                  MIN  LT_  MAX    percent percent percent");
  Serial.print(minLightRight);
  Serial.print("  ");
  Serial.print(sensorValue);
  Serial.print("  ");
  Serial.print(maxLightRight);
  Serial.print("  ");
  Serial.print(adjustedLightRight);
  Serial.print("         ");
  sensorValue = analogRead(sensorLeft)
  if (maxLightLeft < sensorValue) maxLightLeft = sensorValue;
  if (minLightLeft > sensorValue) minLightLeft = sensorValue;
  rangeOfLight = maxLightLeft - minLightLeft;
  adjustedLightLeft = (sensorValue - minLightLeft) /
                          (rangeOfLight / 100);
  Serial.print(minLightLeft);
  Serial.print("  ");
  Serial.print(sensorValue);
  Serial.print("  ");
  Serial.print(maxLightLeft);
  Serial.print("  ");
  Serial.println(adjustedLightLeft);
  delay(500);
}
```

Record your results from the simple light experiment here:

Left Eye	Right Eye
`minLightLeft`	`minLightRight`
`maxLightLeft`	`maxLightRight`
Ambient light value	Ambient light value

Bill of Materials

- Arduino Uno or similar
- 10-kΩ resistor (2) (271-1126) (Radio Shack)
- Cadmium-sulfide (CdS) photoresistor (2) (276-1657) (Radio Shack)
- Parallax Board of Education (35000 or any small prototyping breadboard) (www.parallax.com/BOEshield)

Creating Dynamic Eyes

Now that your android can detect light, we now need the ability to show a pupillary response. Doing this requires eyes. For this experiment, our eyes are 8 × 8 LED matrix displays. We will use two of them and place a photocell on either side. LED matrices normally require multiplexing and use of 16 output lines for a single-color 8 × 8 matrix. To save on pins and make programming simpler, we will use a couple of 8 × 8 bicolor matrix displays from Adafruit. This unit uses an I2C bus to save on input-output (I/O) lines and comes with a full Arduino library. The other benefit of using a module is that it allows for three colors (red, orange, and green) simultaneously. It is also possible to substitute a cheaper one-color display or even a full-color 8 × 8 matrix using the same library. An advanced high-fidelity version of this project could employ small video displays or cell phone displays that show detailed eye images. We are going to keep things simple and inexpensive for this android experiment, however.

Because we are using 8 × 8 LED displays, we can define each row of the display with a byte of binary data. This sort of 8 × 8 image design is a lot like creating custom character sets on old 8-bit computers such as the Atari or Commodore. The characters are easy to design with a spreadsheet, and one is included in the code-file package online so that you can create your own eye designs. The design of each row is to use 1 bit for each LED. We will supply the number to the Adafruit 8 × 8 LED module in the form of eight hexadecimal or decimal numbers because base 16 or base 10 numbers are easier to type than binary numbers (Figure 4-4). The design for this experiment gives the eyes and pupils rounded corners (Figures 4-5 and 4-6). You can design your own eye style and substitute those graphics instead as you wish.

128	64	32	16	8	4	2	1	DEC	HEX
		1	1	1	1			60	3C

Figure 4-4   Creating a binary image. In this example, we want the four middle LEDs of a line to light up. By placing 1s for the lit pixels in this table, all one has to do is add 32 + 16 + 8 + 4, which equals 60. The hexadecimal equivalent is 3C.

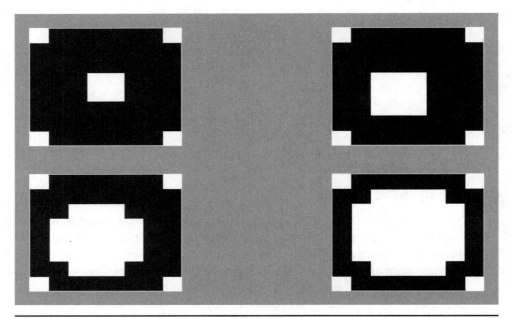

Figure 4-5   Four pupils. Here is the design for a rounded eye that includes four variations on pupil size. The upper right is the default size. Sizes vary from contracted (*upper left*), to normal (*upper right*), to mild dilation (*lower left*), to severe dilation (*lower right*).

Code

Listing 4-2   A single eye using one LED matrix display.

```
/*
Single Dynamic Eye with Matrix LED Display
Androids: Build Your Own Lifelike Robots by Bergeron/Talbot
Listing 4-2
By Thomas Talbot
Arduino 1.0.1 environment
Uses Adafruit I2C LED Backpack Library at www.adafruit.com and
https://github.com/adafruit/Adafruit-LED-Backpack-Library
*/
```

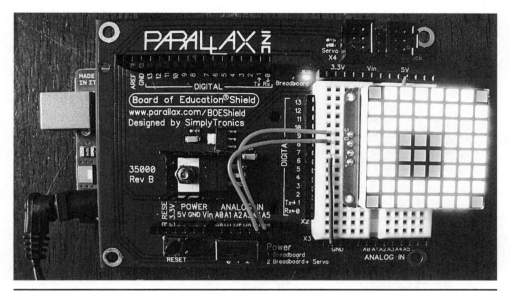

Figure 4-6 8 × 8 LED matrix in action. This example includes animation frames for the pupil of the eye from contracted to fully dilated in four steps.

```
#include <Wire.h>
#include "Adafruit_LEDBackpack.h"
#include "Adafruit_GFX.h"

Adafruit_BicolorMatrix matrix = Adafruit_BicolorMatrix();

void setup() {
  Serial.begin(9600);
  Serial.println("8x8 LED Matrix Test");

  matrix.begin(0x70);  // The default I2C address is 0x70
}

// Bitmaps stored in FLASH memory
static uint8_t __attribute__ ((progmem)) pupil01[]={126, 255, 255,
    231, 231, 255, 255, 126};
static uint8_t __attribute__ ((progmem)) pupil02[]={126, 255, 255,
    199, 199, 199, 255, 126};
static uint8_t __attribute__ ((progmem)) pupil03[]={126, 255, 199,
    131, 131, 131, 199, 126};
static uint8_t __attribute__ ((progmem)) pupil04[]={126, 195, 129,
    129, 129, 129, 195, 126};
```

```
void loop() {

  matrix.setRotation(0);
  // Color values are LED_GREEN, LED_YELLOW, LED_RED

  matrix.clear();
  matrix.drawBitmap(0, 0, pupil01, 8, 8, LED_GREEN);
  matrix.writeDisplay();
  delay(1500);

  matrix.clear();
  matrix.drawBitmap(0, 0, pupil02, 8, 8, LED_GREEN);
  matrix.writeDisplay();
  delay(1500);

  matrix.clear();
  matrix.drawBitmap(0, 0, pupil03, 8, 8, LED_GREEN);
  matrix.writeDisplay();
  delay(1500);

  matrix.clear();
  matrix.drawBitmap(0, 0, pupil04, 8, 8, LED_GREEN);
  matrix.writeDisplay();
  delay(1500);
}
```

Bill of Materials
- Arduino Uno or similar
- Adafruit 8 × 8 bicolor LED matrix display with I2C backpack (www.adafruit.com)
- Parallax Board of Education Shield (35000 or other prototyping board) (www.parallax.com/BOEshield)

A Set of Animated Eyes

Considering that most people have two eyes instead of one, it might be a good idea to add a second I2C LED matrix display. The I2C bus permits up to 128 peripherals to be attached, although each requires its own unique address. Attaching two stock modules to the I2C bus will result in neither working because both displays are set to the same I2C address by default. Fortunately, the displays have the ability to

FIGURE 4-7 Changing the I2C address for a two-eye display. For one of the displays (left eye), solder a wire between the two A0 pads, or make a solder bridge to join the pads. This will change the I2C address from 0x70 to 0x71. Do not solder the pads for the right eye display.

change their I2C address by connecting a pair of solder pads. You can change the address for one of your displays by following Figure 4-7.

In order to have two displays work on the software side, it is necessary to initiate two instances of the matrix controller object called `matrixRight` and `matrixLeft`. One is assigned the 0×70 address, and our modified display is assigned 0×71. The following example will produce a set of eyes. The process for sending an image and updating the matrix display takes three steps: (1) clearing the display, (2) applying the bitmaps, and (3) issuing a `write` command to update the display. For hardware hookup, simply connect the second display to the same lines as the primary display from the preceding example. This code will also work with the "Mark 54 Pupil Examination Simulator" schematic found later in this chapter.

Code

LISTING 4-3 Dynamic eyes with LED matrix displays.

```
/*
Dynamic Eyes with Matrix LED Displays
Androids: Build Your Own Lifelike Robots by Bergeron/Talbot
Listing 4-3
By Thomas Talbot
Arduino 1.0.1 environment
```

```
Uses Adafruit I2C LED Backpack Library at www.adafruit.com and
https://github.com/adafruit/Adafruit-LED-Backpack-Library
*/


#include <Wire.h>
#include "Adafruit_LEDBackpack.h"
#include "Adafruit_GFX.h"


Adafruit_BicolorMatrix matrixRight = Adafruit_BicolorMatrix();
Adafruit_BicolorMatrix matrixLeft = Adafruit_BicolorMatrix();


void setup()
{
  Serial.begin(9600);
  Serial.println("8x8 LED Matrix Test");

  matrixRight.begin(0x70);  // The default I2C address is 0x70
  matrixLeft.begin(0x71);   // Left eye matrix set to 0x71 by
bridging A0 jumper on back
}

// Bitmaps stored in FLASH memory
static uint8_t __attribute__ ((progmem)) pupil02[]={126, 255, 255,
    199, 199, 199, 255, 126};

void loop()
{
  matrixRight.setRotation(3);
  matrixLeft.setRotation(3);
  // Color values are LED_GREEN, LED_YELLOW, LED_RED

  matrixRight.clear();
  matrixRight.drawBitmap(0, 0, pupil02, 8, 8, LED_GREEN);
  matrixRight.writeDisplay();

  matrixLeft.clear();
  matrixLeft.drawBitmap(0, 0, pupil02, 8, 8, LED_YELLOW;
  matrixLeft.writeDisplay();
  delay(15000);
}
```

Animated Eyes That Respond to Light

Now that we have a working set of sensors and a working set of eyes, we can integrate them together to create a working light reflex examination as we build up to constructing a pupil examination simulator. Code Listing 4-4 (online) has the minimal code for integrating the photosensors with animated eyes. We will use some example snippets from that code here in order to explain it.

Unlike the CdS photoresistor calibration program, it is desirable to have the system respond to absolute levels of light. We have no control over whether the simulator is run in a dark room or a bright room, nor do we want the user to have to calibrate our unit every time it runs. This means that you should put your CdS photocell calibration values into your code before running it, as seen in this pseudo code:

```
int maxLightRight = 995;
int minLightRight = 140;
int maxLightLeft = 1000;
int minLightLeft = 540;
```

The basic approach is simply an integration of our sensing formulas—where `adjustedLight` variables are computed to produce two numbers, `adjustedLightRight` and `adjustedLightLeft` that vary between 0 and 120 or so. Because both pupils move in response to input from either eye sensor, the value of light `a` is the higher of the two `adjustedLight` variables:

```
if (adjustedLightLeft > adjustedLightRight) a = adjustedLightLeft
if (adjustedLightRight >= adjustedLightLeft) a = adjustedLightRight;
```

In the pseudocode snippet that follows, the value of light sensed `a` goes through a series of if-then evaluations. Here low values of light less than 20 produce the severe dilation image on both displays. The statements afterward control the other three animations that respond to higher levels of light.

```
if (a < 20)  // severe dilation
   Update display with SEVERE_DILATION bitmap
else if (a < 40) // normal dilation
   Update display with NORMAL_DILATION bitmap
else if (a < 85) // normal pupil
   Update display with NORMAL_PUPIL bitmap
Otherwise a must be > 85   // small pupil
   Update display with MIOSIS_PUPIL bitmap
```

Project: Mark 54 Pupil Examination Simulator

Now that everything works normally from the preceding experiment, let's have some fun breaking the code to simulate abnormal findings. To do this, we need to write code to run a number of checks that will mimic the human examination. First, we have to define what can go wrong:

```
// Pathologies. Change to cause problems
// TRUE means functioning, FALSE means broken
boolean RIGHT_RETINA = true;   // right eye sensor
boolean RIGHT_MOTOR = true;    // right eye pupillary muscle
boolean RIGHT_CHIASM = true;   // right to left communication
boolean LEFT_RETINA = true;    // left eye sensor
boolean LEFT_MOTOR = true;     // left eye pupillary muscle
boolean LEFT_CHIASM = true;    // left to right communication
```

This code, from online Listing 4-5, defines the behavior of three systems for each eye. Any of the preceding systems can be disabled by setting the value to false. First, the retina is the light sensor. Second, the chiasm communicates visual information to the opposite side. And third, the motor pathway innervates the pupillary muscles. The ability to disrupt any of these three things from either side allows for simulating the vast majority of things that can go wrong. Interestingly, left eye data are mostly processed in the right side of the brain, as are muscle outputs. Because of this, the consensual light response in the opposite eye ends up crossing from one side of the brain to the other twice!

After sensing light in both eyes, the code needs to zero out data from any disabled sensors:

```
if (RIGHT_RETINA != true) adjustedLightRight = 0;
if (LEFT_RETINA != true) adjustedLightLeft = 0;
```

Next, communication between brain hemispheres is assessed, and consensual light reflexes can only happen if the chiasm values are true:

```
//  Simulated interhemispheric communication
if ((adjustedLightLeft > adjustedLightRight) && LEFT_CHIASM)
{
  adjustedLightRight = adjustedLightLeft;
}
if ((adjustedLightRight > adjustedLightLeft) && RIGHT_CHIASM)
{
  adjustedLightLeft = adjustedLightRight;
}
```

Finally, we are ready to make sure that the pupillary muscle functions on each side. Here is the display code for the right eye:

```
if (RIGHT_MOTOR == true)
  {
      if (adjustedLightRight < SUPER_DILATED)
      {
          matrixRight.clear();
          matrixRight.drawBitmap(0, 0, pupil04, 8, 8, LED_GREEN);
          matrixRight.writeDisplay();
      }
      else if (adjustedLightRight < NORMAL_DILATED)
      {
          matrixRight.clear();
          matrixRight.drawBitmap(0, 0, pupil03, 8, 8, LED_GREEN);
          matrixRight.writeDisplay();
      }
      else if (adjustedLightRight < NORMAL_PUPIL)
      {
          matrixRight.clear();
          matrixRight.drawBitmap(0, 0, pupil02, 8, 8, LED_GREEN);
          matrixRight.writeDisplay();
      }
      else  // MIOSIS_PUPIL
      {
          matrixRight.clear();
          matrixRight.drawBitmap(0, 0, pupil01, 8, 8, LED_GREEN);
          matrixRight.writeDisplay();
      }
  }
  else
  {   // fully dilated
      matrixRight.clear();
      matrixRight.drawBitmap(0, 0, pupil04, 8, 8, LED_GREEN);
      matrixRight.writeDisplay();
  }
```

When running code Listing 4-5 for this simulator, be sure to put in your own CdS photocell calibration numbers. If the sensitivity of the examination seems off, it is possible to adjust the thresholds for pupil responses by adjusting variables SUPER_DILATED, NORMAL_DILATED, NORMAL_PUPIL, and MIOSIS_PUPIL.

Finally, have fun playing with the simulator (Figures 4-8 and 4-9). Try breaking one or more items by setting the variables to `false` and see how the examination changes. By breaking things (setting items to `false`), the examinations can get interesting. For example, if `LEFT_RETINA` is set to `false`, both pupils will dilate when light is shown into the right eye, but there will be no response when light is shown into the left eye. If you want to construct a more advanced version, connect some switches and write code to change the Boolean variable settings to create a more interactive examination simulator.

Bill of Materials

- Arduino Uno or similar
- Adafruit 8 × 8 bicolor LED matrix display with I2C backpack (2) (www.adafruit.com)
- 10-k Ω resistor (2) (271-1126) (Radio Shack)
- CdS photoresistor (2) (276-1657) (Radio Shack)
- Large prototyping breadboard

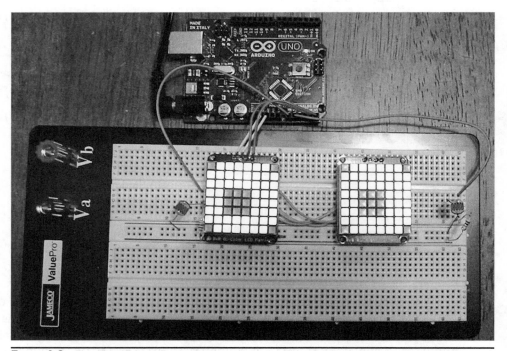

Figure 4-8   The Mark 54 pupil examination simulator. This circuit employs two light sensors and two displays that nicely approximate the eye responses of a human. Changing variables within the code produces pathologies that are identical to those seen in people with visual afflictions.

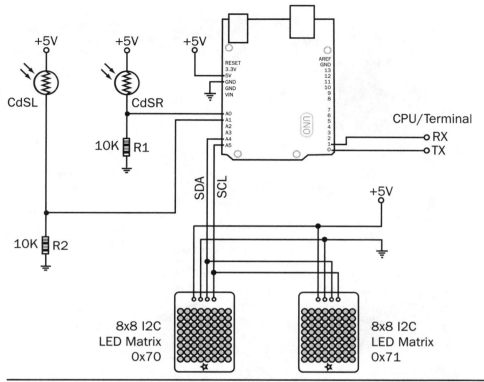

FIGURE 4-9 Schematic for Mark 54 pupil examination simulator. This design combines previous efforts from this chapter. It is possible to run all the code in this chapter up to this point from this design.

Light Tracking and Object Examination

Humans, like most mammals, have eyes that continually move from place to place within the field of vision. Because only a small part of the retina actually obtains high-resolution images, your eyes are always moving and scanning over objects of interest within your field of view. In fact, your eyes are always slewing in rapid motions called *saccades* (Figure 4-10). The human eye scans detailed areas of interest within the visual field, and it is the brain that assembles a complete image. The perception that you, the reader, see a camera-like whole image is a false one—your brain provides the illusion that you see a whole image. Your eyes are scanning over words as you read this page. You probably made several dozen eye movements while reading this paragraph. If you are curious about this phenomenon, try searching reputable sites such as Wikipedia using the term *eye tracking* to learn more. The reason for having eyes move is important; stationary eyes tend to disturb people—people expect some eye movement.

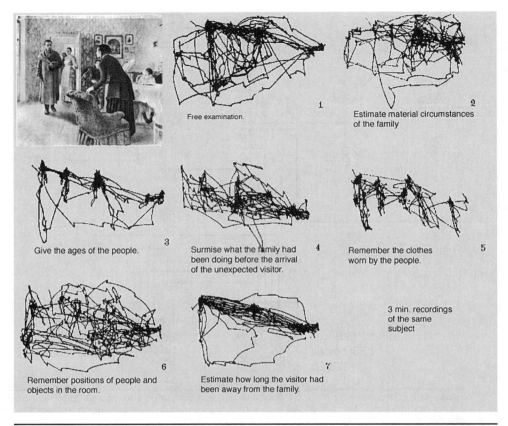

Free examination.   1

Estimate material circumstances of the family   2

Give the ages of the people.   3

Surmise what the family had been doing before the arrival of the unexpected visitor.   4

Remember the clothes worn by the people.   5

Remember positions of people and objects in the room.   6

Estimate how long the visitor had been away from the family.   7

3 min. recordings of the same subject

FIGURE 4-10   Saccades and vision. This illustration shows a picture along with eye-tracking data that show that people scan important parts of an image rather than keeping their eyes still like a camera. The human eye only perceives detail within a narrow field of view. (*Open-source image from www.wikipedia.org under the search term "eye tracking."*)

The Light-Tracking Examination

The light-tracking examination provides a lot of useful diagnostic information to a physician. By having a human move his eyes to follow a flashlight in a wide circle in front of them, it is possible to discern proper function of many systems. The light-tracking examination tests the retina, optic nerve, nerve tracks to the brain, visual processing, motor control, motor tracks, and cranial nerves. An abnormal light-tracking examination can indicate the presence of an injury, a brain tumor, paralyzing conditions, and strokes. Conducting a light-tracking examination is simple: Shine a penlight in front of the face, and then move the light to each side, then up and down, and finally in a big circle. Any normal person or capable android should follow the light with both eyes.

Because the Arduino is too slow to effectively process video images, we will use the CMUcam4 shield, an open-source embedded computer vision processor. CMUcam4, named after Carnegie Mellon University, has an onboard miniature camera, SD-card storage, and a video output for debugging. This device is powered by the Parallax Propeller microcontroller, which employs parallel processing to distribute tasks. Capabilities of the CMUcam4 are image capture to SD-card, histogram generation, automatic tracking of user-defined color blobs, tracked-object image capture, and automatic servo control to aim the unit at tracked objects. Although the CMUcam4 is a large shield, it needs only two wires for serial communication, so stacking it as a shield is not necessary.

An eye-tracking examination is usually performed by having a patient follow a penlight or ophthalmoscope, so we will set the CMUcam4 to track bright-white lights and ignore the rest of its visual field. It is possible to have the CMUcam4 follow objects or lights of a specific color.

The light-tracking information will be read by polling the CMUcam4 over a serial port for coordinates of a light blob, returned as mx and my if a light blob is found (Figures 4-11 and 4-12). Values for mx are 0 for the rightmost field of view, 80 for the center of vision, and 159 for the leftmost field of view. Values for my are 0 for the top, 60 for the center, and 119 for the bottom. To determine what eye animation to associate with the ongoing eye examination, we will split the camera's visual

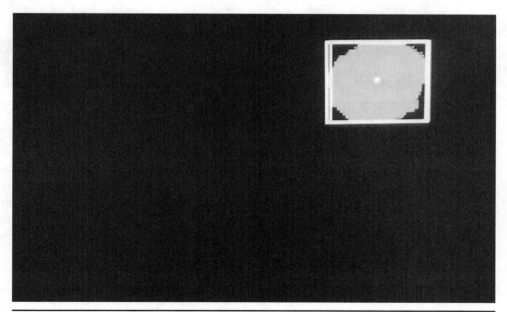

FIGURE 4-11   Video output from the CMUcam4 as it tracks a penlight while running code Listing 4-6. The gray blob on the screen is a light centroid created by shining an ophthalmoscope into the camera. The white point is the center of the centroid, which is surrounded by a bounding box that defines the tracked object. The CMUcam4 reports out the position of the centroid as mx and my over its serial port.

FIGURE 4-12 CMUcam4 is an open-source project with an array of powerful capabilities. This model supports video image output. Note the camera mounted on the surface of the microcontroller.

field into nine areas. The following code is written for the Arduino Mega because it has additional hardware serial ports that allow us to obtain debug output and read-buffered input from the camera. This code can be adapted to run on other Arduinos, but you will lose the debug terminal output on your computer. Hookup is simple: Supply power to the CMUcam4, and connect Serial Port 3 to the TX and RX pins on the side of the CMUcam4.

Code

LISTING 4-6   CMUcam4 light tracking test for Arduino Mega.

```
/*
CMUcam4 light tracking test
```

Androids: Build Your Own Lifelike Robots by Bergeron and Talbot
Listing 4-6. For the Arduino Mega.
*/

```
#include <CMUcam4.h>
#include <CMUcom4.h>

#define RED_MIN 230
#define RED_MAX 255
#define GREEN_MIN 230
#define GREEN_MAX 255
#define BLUE_MIN 230
#define BLUE_MAX 255
#define LED_BLINK 5 // 5 Hz
#define PIXELS_THRESHOLD 5
#define CONFIDENCE_THRESHOLD 50   this 0=0 percent - 255=100
percent.
#define NOISE_FILTER_LEVEL 2

CMUcam4 cam(CMUCOM4_SERIAL3);   // For Arduino Mega only
// use the CMUCOM4_SERIAL if not using Arduino Mega.

void setup()
{
  Serial.begin(9600);
  cam.begin();
  cam.LEDOn(LED_BLINK);
  delay(5000);
  cam.autoGainControl(false);
  cam.autoWhiteBalance(false);
  cam.LEDOn(CMUCAM4_LED_ON);
  cam.colorTracking(false);
  cam.noiseFilter(NOISE_FILTER_LEVEL);
}

void loop()
{
  CMUcam4_tracking_data_t data;
  cam.trackColor(RED_MIN, RED_MAX, GREEN_MIN, GREEN_MAX, BLUE_MIN,
                 BLUE_MAX);
  for(;;)
  {
```

```
   cam.getTypeTDataPacket(&data);
   if(data.pixels > PIXELS_THRESHOLD)
   {
     if(data.confidence > CONFIDENCE_THRESHOLD)
     {
       Serial.print(data.mx);
       Serial.print("      ");
       Serial.println(data.my);
     }
   }
   else
   {
     Serial.println("I don't see anything");
   }
  }
}
```

Interpreting the Data from CMUcam4

Because our goal is to have the eyes looking at where the light is shining, we can start with nine different bitmaps for the eye to depict the directions the eye can look (Figure 4-13).

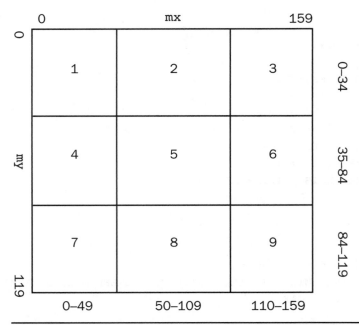

FIGURE 4-13   The eyes look in nine directions. These bitmaps will be used for our light-tracking simulation.

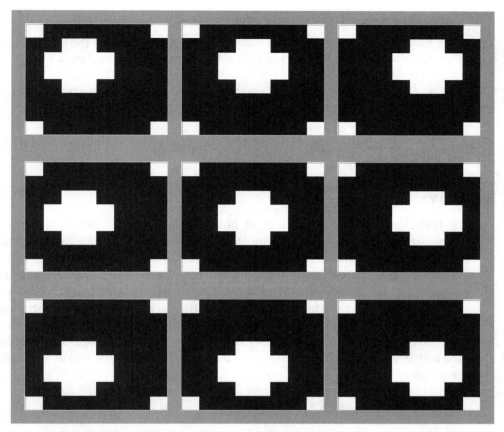

FIGURE 4-14   The CMUcam4 field of view is divided here into nine segments. Point of reference is that of the camera, so area 1 would represent the upper-left field of view, and area 9 would represent the lower right. The center areas are taller and wider than the peripheral areas to maintain a larger center of vision.

FIGURE 4-15   This graphic shows what an examination looks like when a light is present in the android's upper-right field of view.

Bill of Materials

- Arduino Mega or similar with multiple hardware serial ports
- Adafruit 8 × 8 bicolor LED matrix display with I2C backpack (2) (www.adafruit.com)
- CMUcam4 (available at www.robotshop.com or www.sparkfun.com)
- Large prototyping breadboard

Optionally, use the circuit for the Mark 54 pupil examination simulator and add the CMUcam4 to it.

Summary

In this chapter we have looked at how to make your android gather visual information from the world and how to use eyes as a method of communicating to others that the android has this information. Besides making accurate physical examinations for medical simulation purposes (Figure 4-16), you can tailor these features to track a person moving around the room and have your android's eyes follow that person. Even better, you can tailor the tracker to follow a person wearing a certain color shirt so that the eyes only follow the person wearing that color. You also made your android appear very human by simulating a pupillary response to the ambient lighting conditions. This is a thoughtful detail that makes any android seem more lifelike.

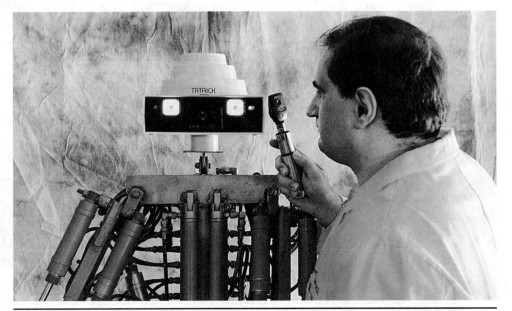

Figure 4-16   Both the authors are physicians. Here Thomas Talbot is conducting a pupillary examination on his android, Tatrick.

CHAPTER **5**

All Ears

Our ability to localize sounds quickly and accurately is key to survival. Identifying the general location of the rattle of a rattlesnake preparing to strike can be just as important as localizing the beep of a delivery van backing into your car. In both cases, where milliseconds count, you can take action before the threat is in your visual field.

The ability to localize sounds also contributes to our social fabric. It's appropriate to face someone as they address you, for example—just don't eyeball them if they're your military superior. However, the process of sound localization is computationally intensive, especially when the noise level is high. To cope, we have developed extensive biophysical signal-processing capabilities that involve both amplitude and temporal calculations.

In this chapter we'll explore the basics of sound localization and associated behaviors with a few inexpensive electret microphones, passive infrared (PIR) motion detectors, a servo-controlled turret, a pair of optoisolators, light-emitting diodes (LEDs) and other indicators, and an Arduino microcontroller. We'll also look at several modifications and review some practical applications of the technology to your android design.

Biological Basis

The localization of sound may seem trivial compared with the task of sound recognition, much less speech recognition and interpretation. However, it remains as one of the most complex processing problems handled by our massive brains. Furthermore, not only is sound localization a significant ability in its own right, but it also supports the recognition and interpretation of sounds. To illustrate, consider the following scenario.

Phoenix Rises

As you rub the dust from your eyes, you realize that it's too dark to see anything. You feel the weight of rubble on your thighs, pinning you down. As you gain a sense of what happened—you were at your desk one minute and covered with rubble the next—obviously an earthquake has occurred or perhaps something has crashed into your office building.

Then you hear a slight whirr and thuds of what sounds like cinder blocks being tossed to the side against a background of creaking and occasional crashes of what must be floors collapsing onto one another. You yell for help for what seems like forever until you suddenly notice a rescue android by your side. It injects you with something and applies a tourniquet before tossing the rubble off your crushed leg and then scoops you up. Once clear of the building, human medics give you another injection to prevent the toxins from the crush injury from stopping your heart.

Sounds simple enough, right? Even setting aside the challenge of diagnosing the situation and administering the appropriate medical treatment, the task before the android was all but simple. Amid the sounds of crumbling rubble, the android recognized your cry for help and localized the sound before crawling to your location. It may also have relied on vibration sensors, IR sensors, and perhaps even a carbon dioxide or other gas detector to determine your location. But sound-based localization has the advantages of extended range—sounds travel hundreds of meters—and the ability to penetrate materials that are opaque to IR and visible light.

A Bit of Biophysics

Sound, the interpretation of audio frequency vibrations by the human auditory system, is a complex sensation that has a basis in neurology, anatomy, and human perception. In the following sections we'll review some basic biophysical concepts as well as the sensations of sound localization and the related topics of audio imaging and psychoacoustics.

Speed of Propagation

The propagation of audio-frequency vibrations in the air is a relatively simple physical phenomenon, consisting of longitudinal waves of compression emanating from the source, as illustrated in Figure 5-1. These acoustic vibrations travel at 343.2 m/s (1,126 ft/s) in air at 20°C (68°F). Assuming that the external ear openings are about 18 cm (7 in.) apart, waves of compression emanating from a listener's left will reach the frequency-sensitive hair cells in the cochlea of the listener's right ear about 0.50 ms after reaching the corresponding sensors in the listener's left ear.

Also note the acoustic shadow depicted in Figure 5-1. This shadow or area of attenuated waves of compression results from absorption and reflection of the waves by the listener's head. As a result of the shadow, the amplitude of the

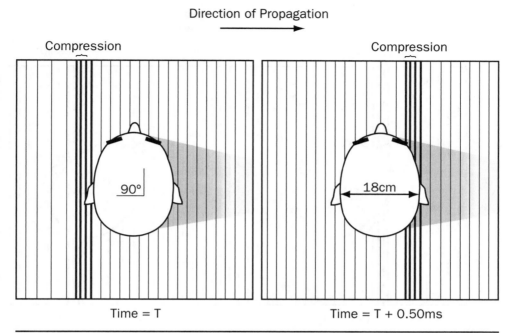

FIGURE 5-1 Acoustic compression wave at time *t* (*left*) and time *t* + 0.50 ms (*right*). Source to listener's left. Note the acoustic shadow to the listener's right.

compression wave reaching the listener's right ear will be diminished significantly from the amplitude of the wave directly in line with the listener's left ear.

Longitudinal waves of compression emanating from a source directly in front of the listener will reach both ears simultaneously and without amplitude changes due to a shadow effect. And between the two extremes, the compression waves will reach the right ear somewhere between 0 and 0.50 ms after reaching the left ear, depending on the relative orientation of the listener's ears and the source. For example, if the listener is facing the source at a 45-degree angle, as in Figure 5-2, the distance between the left and right ears, relative to the source, is about 13 cm (i.e., $c^2 = a^2 + b^2$). The acoustic wave would require about 0.36 ms to travel from the left to the right ear.

In addition, note the change in the acoustic shadow, in that the listener's right ear is no longer in the center of the shadow. Also, the listener's left ear is no longer directly in line with the pressure wave. As a result, the difference between the amplitude of the pressure waves at the left and right ears is diminished. This brings us to the next topic—intensity.

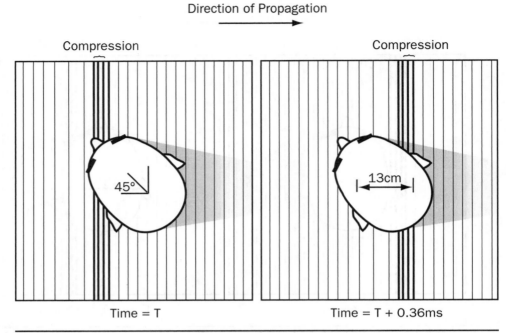

Direction of Propagation

Compression   Compression

45° |←13cm→|

Time = T   Time = T + 0.36ms

Figure 5-2   Acoustic compression wave at time *t* (*left*) and time *t* + 0.36 ms (*right*). Source between left and center. Ambient = 20°C (68°F).

Intensity

The *intensity* of acoustic vibrations is usually quantified in decibels (dB) according to the following formula:

$$dB = 10 \log I_1/I_2$$

where I_1 is the intensity or power of the first acoustic vibration, and I_2 is the intensity of the second acoustic vibration. From this formula, you can see that if I_1 is 10 times greater than I_2, the intensities differ by 10 dB. Similarly, if I_1 is twice the intensity of I_2, the difference is 3 dB. Note that the ratio measurement expressed in decibels is dimensionless. When human hearing is involved, it's common to substitute the lowest intensity of acoustic vibration that we can perceive—about 10^{-13} W / m² —for I_2 in the formula. The result is an absolute scale that reflects the human sense of hearing.

Regardless of whether the intensity measure is relative or absolute, there are certain properties worth mentioning. For one, our sense of loudness is nonlinear. If you crank up your stereo from 2 to 4 W, you'll double the acoustic intensity. However, if you crank your stereo from 20 to 22 W, you probably won't detect a difference (recall the *just noticeable difference* from Chapter 2). Instead, you'll have to increase power by 20 W, for a total of 40 W, to perceive the same relative increase in volume. As such, describing acoustic intensity in terms of decibels is a natural.

Another property of acoustic intensity is that it diminishes with the square of distance from the source. Double the distance, and the intensity of acoustic vibration is diminished by a factor of 4. Triple the distance, and the intensity is diminished by a factor of 9.

When it comes to quantifying acoustic waves, intensity or power isn't necessarily the best metric. Because ordinary microphones respond to sound pressure level (SPL) as opposed to the power of the source, it's more common to measure SPL than intensity. The formula for SPL ratios in decibels is

$$dB = 20 \log SPL_1 / SPL_2$$

where SPL_1 is the sound pressure level of the first acoustic vibration, and SPL_2 is the sound pressure level of the second vibration—typically the reference. The significance of this formula is that the ratio doubles every 6 dB instead of every 3 dB. Think of SPL as the equivalent of voltage in an electrical circuit. This difference between SPL and acoustic intensity or power is important to keep in mind when you interpret acoustic meter readings and microphone specifications.

Sound Localization

Sound localization is the ability of the auditory system to determine the physical location of a sound source by auditory cues. For example, imagine that you are standing in an open field, and off in the distance and to your left, a multidimensional portal opens, creating tremendous vibrations as the air is sheared at a molecular level. Even with your eyes closed, you should be able to identify, within about 1 degree of arc, the direction of the horrendously loud phenomenon.

Not only would the acoustic vibrations reach your left ear first, but the amplitude of the vibrations reaching your left ear would be greater than those reaching your right ear in part because of the shadowing effect of your head. That is, your dense, fluid- and tissue-packed head is between the audio source and your right ear, whereas your left ear is in direct line with the source. Sound localization, then, is at least partially a function of both relative time delay and relative amplitude of the sound pressure wave—that is, the sound pressure level.

In addition to the significant contributions of relative amplitude and phase of acoustic vibrations, localization is also a function of the directional characteristics of the external ears (pinnae), which modify the vibrations reaching each ear. The external ears are predominantly useful for localizing the source of audio frequencies less than about 6 kHz.

The characteristics of the acoustic vibrations affect the ability of our ears and associated nervous system to localize sounds. For example, the human auditory system is relatively insensitive to differences in the arrival time or phase of pure tones at frequencies above 6 kHz. Think about the warble of a police siren—it's everything but a pure tone, and it's well below 1 kHz. The 6-kHz figure is significant because it covers the frequency spectrum of human speech as well as the calls made by wild animals.

Another factor that contributes to sound localization is the equivalent of sensor fusion from multiple sense organs. Auditory cues are combined with information from the position and movement sense organ in the ears (the *vestibular system*), eyes, and motion and position sense organs in the muscles, tendons, and joints (the *kinesthetic system*). To get a sense of this sensor fusion in action, consider the automatic reflex action of rotating the head from side to side to better localize the source of a sound. The resulting variation in the relative amplitude and phase relationships of signals reaching the ears provides the auditory system with additional data points that are used to more accurately localize the signal source.

Audio Imaging

A discussion of sound localization would be incomplete without mentioning audio imaging. Whereas the term *sound localization* is used commonly to describe the perceived position of an audio source in a flat or two-dimensional (2D) sense, *audio imaging* refers to the more complex, three-dimensional (3D) perception of a sound source. An example of audio imaging is the sensation you experience when listing to a stereophonic recording through a stereo sound system. You can probably imagine or deduce the relative positions of each instrument and the vocalist, say, drums center stage, vocalist right stage, and piano left center stage. The auditory system produces an audio image of the sound source in a manner much akin to the way our binocular visual system processes visual signals to create 3D visual images.

If you're going to create an android that can accurately pass a human hearing test, then you'll have to handle both intra-aural sounds provided by headphones and free-space or external sources. Perceptions of the two sources differ in part because the sound pressure wave interactions that occur in free space are not found in intra-aural systems. In addition, the previously described shadowing effect of the head, as well as the directional characteristics of the external ears, don't contribute to your interpretation of sounds from intra-aural sources.

The predominant perceptual difference between intra-aural and free-space systems is where the audio image is formed. If you listen to a stereo recording through loudspeakers, the audio image should be formed outside your head. However, if you listen to the same music through headphones, the sounds should appear to emanate from within your head, localized on an imaginary line drawn between both ears. This lateralization of sound, which is a function of relative intra-aural phase differences, is the basis for standard hearing tests.

Psychoacoustics

Several properties of the human auditory system defy explanation on a strictly physiologic or anatomic basis but are instead best understood in terms of human perception of sound, or *psychoacoustics*. The psychoacoustic property most applicable to our discussion of localization is *perceived intensity*. The perceived intensity of a sound is a function of the audio signal's duration. Whereas sounds that last longer than about 250 ms and are of equal amplitude are perceived as having equal intensity; shorter-duration sounds of the same amplitude are perceived to have a lower intensity.

Think of a car horn. A short toot doesn't seem as loud—or obnoxious—as the sound of someone leaning on the horn for 10 seconds or more. And a 10-second blast seems just as loud as a 6-second blast. The short toot just doesn't sound as loud, even though it's from the same horn as the longer blasts. Quantitatively, a decade increase in duration, say, from 50 to 500 ms, is equivalent to a 10-dB increase in intensity—as long as it involves crossing the 250-ms threshold.

Another psychoacoustic property is that, through conditioning, some sounds are pleasant and others are annoying. Take the flat-5(bV) note, which is commonly a component of the audio warble blasted into city streets by police and ambulance sirens. With the exception of blues, the flat-5 note is rarely used in modern Western music. Even then, the note is used only as a passing tone, meaning as a brief tone between two other notes.

Normal Behavior

Regardless of how sound is localized, it's the application of localization data that counts. So how does sound localization translate into observable behavior? Well, if you've seen the demonstrations of motorized camera platforms following a red ball or other colored object, you've seen what human's don't do. If you tracked every sound by turning your head, you'd have arthritis in your neck by your early twenties. Instead, we use our eyes to track the sources of significant sounds in the environment and save the slower head movements for situations that demand more immediate attention.

Relevance to Android Designs

As with humans, an android's ability to localize sounds can be key to both survival and to fitting in with humans. Sales of future servant androids will no doubt depend heavily on personable, efficient machines. And today, there's utility in providing a robot with the means to track or follow an audio signal.

Of course, there's the higher-level function of sound recognition, including speech recognition, that's beyond the scope of this chapter. However, sound localization and recognition are not an either-or proposition. As shown in Figure 5-3, localization and recognition can be implemented as independent, parallel

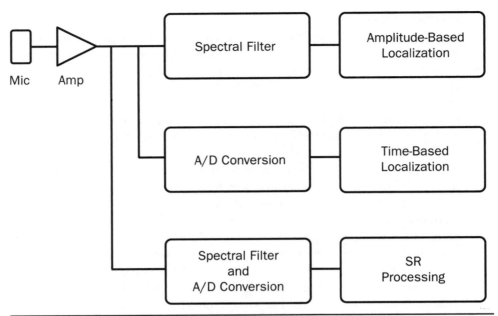

FIGURE 5-3 Simultaneous processing of audio for recognition and localization.

processes. As the audio is captured, it can be processed in parallel for amplitude, time of arrival, and signal content.

Small-vocabulary, discrete speech-recognition shields for the Arduino and other microcontrollers are available and potentially useful, especially if you use a dedicated processor in a network configuration. On the other hand, large-vocabulary, continuous speech recognition is beyond the capabilities of an Arduino, but data extracted from the sound can be relayed to a desktop computer or server for processing.

Sound-Localization Experiments

Compared with previous chapters, the experiments described here require significant finesse in working with hardware. It may need a bit of tweaking to get the systems described here to operate as advertised. This is due in part to the analog nature of the sensors and amplifiers and in part to the unknown elements in your experimental environment. For example, your system may work perfectly in an acoustically dead room with carpet and heavy drapes and fail miserably when used in an echo chamber that doubles as a tile-lined bathroom.

The goal in the series of experiments defined herein is to expose you to a variety of approaches to sound localization. We start with relatively simple amplitude-based localization using the eyes and then the eyes and neck rotation to indicate response to a sound. Then we introduce the computationally intensive time-based approach to localization, including the use of high-speed hardware and direct port manipulation. We finish with a few suggestions on alternatives to using discrete light-emitting diodes (LEDs) in each eye to indicate response to sound.

As you work through these experiments, bear in mind that this is by no means an exhaustive review of the technologies available for sound localization. There are specific sensors designed for sound localization and special high-speed hardware optimized for the task. However, these solutions start at several thousand dollars—presumably beyond what you have budgeted for this aspect of your android design.

Amplitude-Based Sound Localization

In this first experiment, we'll use the relative amplitude of an audio signal, as detected by a pair of electret microphones, to determine the general location of the source. The measurement is crude by human standards—tens of degrees versus 1 second of arc—but sufficient to illustrate the principles. We'll use eye movement as a feedback mechanism for sound localization, represented by three LEDs per "eye."

Bill of Materials

To construct the amplitude-based sound-localization system, you'll need the components listed below and shown in Figure 5-4:

- Electret microphones with amplifiers (3)
- Discrete LEDs (6)
- Breadboard
- Arduino Uno
- 5-V direct-current (dc) power supply
- Jumpers or wires
- Acoustic signal source
- 500-Ω, ¼-W resistor
- Head-sized support structure
- Acoustically dead environment (or nearly dead)
- Prototyping shield (optional)
- SPL meter (optional)

The acoustic signal source in the figure is a self-powered studio monitor that's designed for relatively flat response from a few hundred hertz to 20 kHz. The self-powered woofer normally paired with the monitor is intentionally not used in this experiment because of the relative nondirectivity of low-frequency audio. We've provided a set of 30-second MP3 files online, with audio tones at 500 Hz and 1, 2, 3, 4, and 5 kHz for use in this experiment. No studio monitor or MP3 player? No problem. The high-frequency "click click" from an ordinary retractable ink pen works fine for distances up to a few feet from the microphones. The cheaper pens tend to create a louder series of clicks.

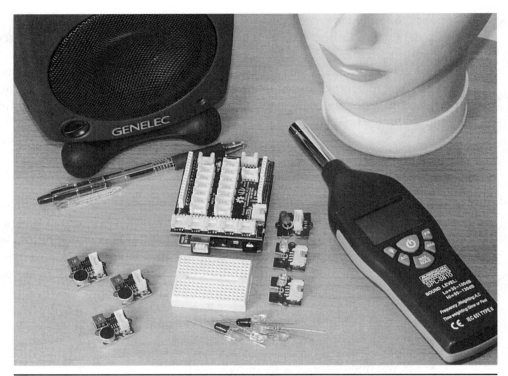

FIGURE 5-4   Main components of the authors' amplitude-based localization experimental setup.

The most critical element in this setup is the set of three electret microphones with amplifiers. There are dozens of sources of microphone elements and small audio amplifiers on the market. Although separate microphones and amplifiers will work, integrated amplifiers are easier to set up and use. If you're using the Grove prototyping shield described in previous chapters, then the electret microphones with integrated LM386 amplifiers from Grove are a reasonable choice because they're compact, and the gain can be adjusted with an onboard potentiometer. The downside of the Grove module is that the components are surface-mount, meaning that it's difficult to modify the circuit to provide, for example, filtering of the audio. In addition, the electret microphones are omnidirectional, and directional is best, but not absolutely necessary, for this application—more on this later.

If you can't find a set of amplified microphones that you like, then consider building the microphone using the schematic shown in Figure 5-5. The resistors are ¼ W, and the capacitors are rated at 12 V or more, assuming a 5-Vdc supply. An advantage of the LM386 over many other operational amplifiers is the ability to run on a single-sided 5-Vdc power supply. Many operational-amplifier chips require positive and negative power supplies.

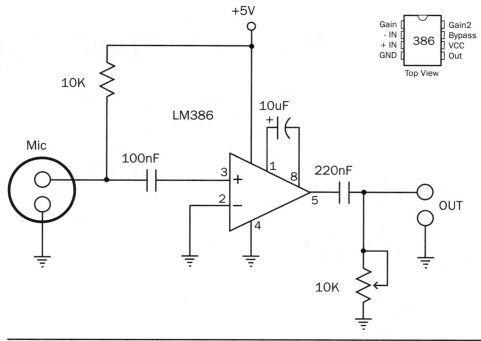

Figure 5-5 Schematic of an electret microphone with integrated LM386-based amplifier.

When you're building out the amp, pay particular attention to the polarity of the electret microphone element. The case is normally connected to ground. Also, if you have a choice, go with directional electret microphone elements.

As noted earlier, the localization system works best in an acoustically dead (i.e., sound-absorbing) room. Operate the system away from walls and windows to minimize reflections. Also, you'll get better results if you minimize extraneous noise, such as sound from your stereo, video game, or computer.

Although not necessary, an inexpensive SPL meter is a handy tool to have for measuring the ambient noise level and the level of the sound source. You might be surprised at the difference between daytime and nighttime noise levels, especially if you live near a major highway. An SPL meter is also handy for quantifying the shadow effect of your support structure and effectiveness of any sound barriers you might have in place to limit extraneous noise.

Circuit

The circuit for the amplitude-based localization circuit, shown in Figure 5-6, consists of three amplified electret microphone modules and a half-dozen discrete LEDs. The microphone modules are designated MicL, MicR, and MicC and are intended for placement left, right, and center, respectively, relative to each other. Note that the modules feed analog ports on the Arduino.

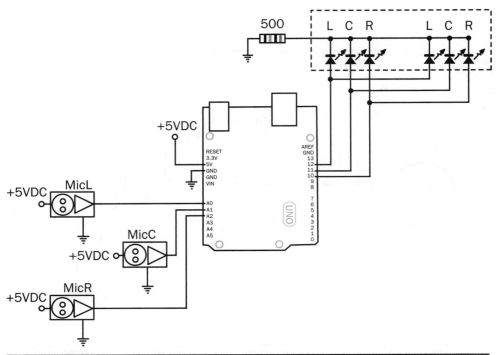

FIGURE 5-6   Schematic of the amplitude-based sound-localization circuit.

The six discrete LEDs are used to indicate eye movement to the left, right, or center, depending on the localization of incoming sounds. The 500-Ω resistor limits the current supplied by the Arduino's digital out pins. Recall that the Arduino is rated for 40 mA per output pin.

Construction

The critical component of construction consists of positioning the three microphone modules in a triangle with MicC facing forward, and MicL and MicR facing left and right, respectively. Start off with a spacing of 7 in. between each module, the approximate distance between a male adult's external ear canals. Increase the spacing if you have trouble obtaining consistent results.

The supporting structure for the three microphones should absorb sound. A foam-filled latex mannequin head used to hold hats and wigs works nicely and serves as a reminder that we're working on androids. As shown in Figure 5-7, a sweatband is a great scaffold for the microphone modules. Cables run down the back of the head to the Arduino and to the breadboard with the LED "eyes." If you don't have a mannequin head handy, consider an inexpensive foam exercise roller. If that doesn't work out, your options include a small watermelon or cantaloupe or simply a roll of paper towels (in a stand for stability). You can even wear the

FIGURE 5-7 Microphone modules mounted on a foam-filled latex mannequin head.

sweatband with microphone sensors. The voltage is low enough that your chances of electrocution are minimal.

A minor task is mounting the two sets of LED "eyes." You can mount the LEDs and resistor on a small breadboard and place the board in front of the support structure. Alternatively, you can mount the LEDs on a circuit board and embed the board in the mannequin, in a cavity dug out of what would be the eye socket.

Code

The Arduino code for amplitude-based sound localization is shown in Listing 5-1. Overall, the program simply performs an analog read on the three microphone modules in the sequence left, center, right. The relative amplitudes of the signals produced by the three modules, as measured by the Arduino's analog-to-digital (A/D) converters, are used to localize the source of the sound.

LISTING 5-1 Arduino code for amplitude-based sound localization.

```
/*
Amplitude-Based Sound Localization
Androids: Build Your Own Lifelike Robots by Bergeron and Talbot
```

```
Listing 5-1. See www.mhprofessional.com/Androids for fully
documented code
Arduino 1.0.1 environment
*/

const int MicLeftPin = A0;
const int MicCenterPin = A1;
const int MicRightPin = A2;
const int MicTrigger = 500;
const int IndexSteps = 3;
const int LEDEyeRightPin = 10;
const int LEDEyeCenterPin = 11;
const int LEDEyeLeftPin = 12;
const long glanceMillis = 500;

int MicLeftVal = 0;
int MicCenterVal = 0;
int MicRightVal = 0;
int MicDirectionVal = 0;
int MicDirectionIndex = 0;
long lastMillis = 0;
unsigned long currentMillis = millis();

void setup() {
  pinMode(LEDEyeRightPin,OUTPUT);
  pinMode(LEDEyeCenterPin,OUTPUT);
  pinMode(LEDEyeLeftPin,OUTPUT);
  EyesCenter();
}

void loop(){
  MicLeftVal = analogRead(MicLeftPin);
  MicCenterVal = analogRead(MicCenterPin);
  MicRightVal = analogRead(MicRightPin);
  currentMillis = millis();
  if(MicLeftVal >= MicTrigger || MicRightVal >= MicTrigger){
    MicDirectionVal = MicRightVal - MicLeftVal;
    if(MicDirectionVal >= 0)
    MicDirectionIndex = IndexSteps + ((MicDirectionVal *
IndexSteps) / MicRightVal);
    else
```

```
    MicDirectionIndex = IndexSteps + ((MicDirectionVal * IndexSteps)
        / MicLeftVal);
if (currentMillis - lastMillis > glanceMillis){
reset();
if (MicDirectionIndex >=4)
EyesRight();
else if  (MicDirectionIndex <=2)
EyesLeft();
}
  }
else
if (currentMillis - lastMillis > glanceMillis){
  if (MicCenterVal >= MicTrigger)
  EyesFlash();
  else
  EyesCenter();
}
}

/*

  --------------------------------------------------------
  EyesCenter()
  EyesLeft()
  EyesRight()
  EyesFlash()
  reset()
  --------------------------------------------------------
  */

void EyesCenter() {
digitalWrite(LEDEyeRightPin, LOW);
digitalWrite(LEDEyeCenterPin, HIGH);
digitalWrite(LEDEyeLeftPin, LOW);
}

void EyesLeft() {
digitalWrite(LEDEyeRightPin, LOW);
digitalWrite(LEDEyeCenterPin, LOW);
digitalWrite(LEDEyeLeftPin, HIGH);
}

void EyesRight() {
```

```
digitalWrite(LEDEyeRightPin, HIGH);
digitalWrite(LEDEyeCenterPin, LOW);
digitalWrite(LEDEyeLeftPin, LOW);
}

void EyesFlash() {
  digitalWrite(LEDEyeRightPin, LOW);
  digitalWrite(LEDEyeLeftPin, LOW);
  digitalWrite(LEDEyeCenterPin, LOW);
  delay (200);
  digitalWrite(LEDEyeCenterPin, HIGH);
  delay (200);
  digitalWrite(LEDEyeCenterPin, LOW);
  delay (200);
  digitalWrite(LEDEyeCenterPin, HIGH);
}

void reset() {
   lastMillis = currentMillis;
}
```

Working from top to bottom of the listing, in the constant declaration section, note the glanceMillis constant, which defines the minimum duration, in milliseconds, that the LED "eyes" will glance toward a sound once triggered. If you shorten the duration, it may be difficult to track the eyes depending on the brightness of your LEDs. If you lengthen the duration, the visual feedback system is unavailable for new signals for longer periods of time.

As in most Arduino programs, the main event is the loop. After the three analog pins are read sequentially, the elapsed-time clock is zeroed as the currentMillis variable is set to the current milliseconds millis().

If either the right or left microphone results in a signal greater than or equal to the trigger value, the values returned by the A/D converter in the Arduino are compared. The greater the difference between left and right microphone values, the further the sound source is from midline. Note that the difference between the left and right microphone values is indexed so that small changes in LED positioning can be achieved, given a rich enough display system. If you increase the value of IndexSteps, you increase the theoretical resolution of the eye-based display system. However, as you increase the value of IndexSteps, differences between steps may be due to noise.

Toward the bottom of the loop, the signal from the center microphone is evaluated relative to its trigger level. If the threshold is exceeded, the eyes are flashed, Goa'uld style, indicating that a sound was detected and that it is midline.

Without the flash, it's impossible to tell whether the system is nonresponsive or that a sound was detected midline.

The subroutines deal with the visual feedback system that relies on the LED "eyes." As discussed in the "Modifications" section, this opens up possibilities for easy substitution with more advanced feedback options.

Operation

In operation, simply click your pen to the left or right of the mannequin head or other support platform and look to the LEDs for feedback. If your system fails to distinguish direction, then remove the microphones from the support, and increase the separation to a foot or more, adding cable to reach the Arduino if necessary.

If the direction seems off, then check the gain controls on each microphone amplifier. Gain should be about equal on each amplifier, but this doesn't mean that the potentiometers are set to the same position. Variations in analog devices, including the LM386, mean that every analog amplifier circuit is different. You can verify equal gain by using the Arduino Uno to print out the values returned by the three A/D converters connected to a microphone.

Modifications

Sound localization is such a rich topic that we could devote the entire book to modifications. But we've constrained ourselves to a few simple mods. First, we'll enrich the feedback indicator and illustrate a practical application of the technology by adding a pair of PIR sensors and a rotating neck. Next, we'll increase the localization accuracy of the system by upgrading the electret microphone elements and enhancing the support structure. Then we'll add a psychoacoustic response so that the system responds to lower-amplitude, long-duration sound. For the adventuresome, we'll switch to a time-based location scheme using relative arrival time, not amplitude, to determine location. Then we'll look at ways to increase the accuracy of the system by moving to a faster processor and by using direct port manipulation on the Uno. We'll finish off with some alternatives to the discrete LED "eyes," including some easy-to-use monochrome and color LED matrices.

Additional Sensors and Feedback

The goal of this first modification is to supplement the uninspiring LED "eyes" with some physical feedback and to show how data from passive infrared (PIR) sensors can be fused with the acoustic data to improve the response of the overall system. Just as we tend to follow noise-producing objects in the environment with our eyes and turn our heads only to focus on a potential threat or when something or someone invades our personal space, this mod will result in a head that rotates when both the PIR and sound sensors agree.

Bill of Materials

To construct this next phase of the project, you'll need the items from the original experiment plus the following:

- PIR sensors (2)
- Pan or turret
- Hobby servo

Figure 5-8 shows the new assembly of suggested parts. The critical components of this modification are the two PIR sensors. We use the PIR modules from Grove here because we're using the Grove shield and because the digital output hold time can be adjusted with an onboard potentiometer. Otherwise, we'd have to adjust the hold time after triggering in software—a minor inconvenience. If you're not constrained by a shield system, then consider the Zilog ePIR, which is both affordable and incredibly powerful. This miniature device has adjustable sensitivity, hold time, and ambient light threshold.

You could get away with a pair of LEDs to indicate motion of the neck to the right or left. However, if you have access to a servo-controlled pan, the motion adds to the realism of the feedback. We had good luck with a plastic ServoCity DPP155

FIGURE 5-8   Additional components required for the first modification.

Pan Kit with an HS-475HB servo. If your budget allows, you might consider a higher-quality pan made of aluminum or pressed steel. The downside of a hollow plastic turret, compared with a metal unit, is a higher noise level during operation of the servo. And, in this application, noise is our enemy.

Circuit

The schematic for this first mod is shown in Figure 5-9. The circuit consists of the original microphone modules and discrete LED "eyes" with the addition of the turret servo and a pair of PIR sensors, PIRL for the left and PIRR for the right. The servo and PIR sensors each require a digital input-output (I/O) port on the Arduino. Consider using an external 5-Vdc supply, as shown in the figure, instead of the onboard regulator. Under load, the servo could draw more current than supported by the regulator.

Construction

Mount the structure from the first part of this experiment onto the pan, as shown in Figure 5-10. Depending on the design of your pan, attach the PIR sensors to the pan or to the base of the structure supporting the microphone assemblies. Angle the PIR sensors so that their coverage nearly—but not quite—overlaps at the front

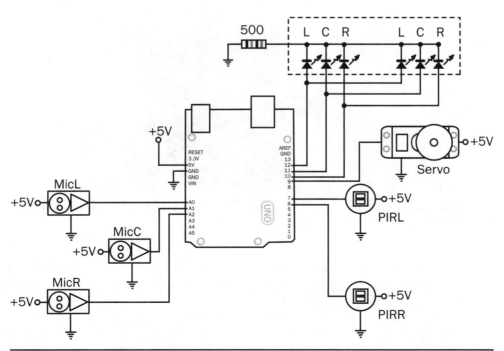

Figure 5-9 Schematic of the amplitude-based sound-localization circuit with PIR sensors and neck servo motion.

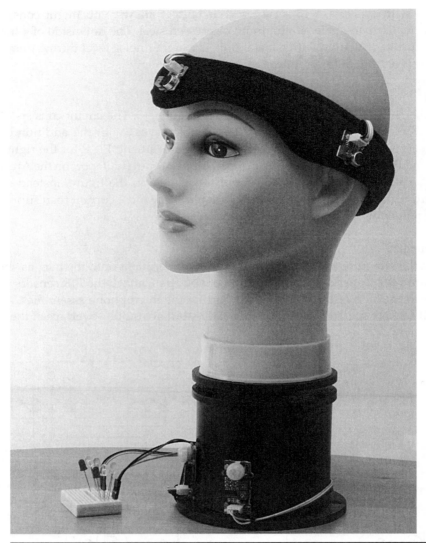

FIGURE 5-10   Modification with PIRs mounted at the base of the servo-controlled pan. Note the two pairs of LED "eyes" near the base of the pan.

of the structure. If you use the Grove PIR sensors, coverage is about 60 degrees for each sensor, or 30 degrees to either side of center. As such, the sensors should be mounted facing 2 × 30 or 60 degrees apart.

If, during testing, you find that the localization accuracy is lower than in the original design, then vibration and noise from the servo are probably at fault. Try adding a layer of foam rubber or similar material between the base of the mannequin or other platform and the pan.

Code

The Arduino code for the modification with additional sensors and feedback, shown in Listing 5-2, leverages the standard servo library. This should be familiar if you've worked through the first two chapters.

LISTING 5-2   Arduino code for sound localization with addition of PIR sensors and neck servo for feedback.

```
/*
Amplitude-Based Sound Localization with PIR Sensors and Neck Servo
Androids: Build Your Own Lifelike Robots by Bergeron and Talbot
Listing 5-2. See www.mhprofessional.com/Androids for fully
documented code
Arduino 1.0.1 environment
*/

#include <Servo.h>
Servo NeckServo;

const int MicLeftPin = A0;
const int MicCenterPin = A1;
const int MicRightPin = A2;
const int MicTrigger = 500;
const int IndexSteps = 3;
const int PIRRightPin = 6;
const int PIRLeftPin = 7;
const int NeckServoPin = 9;
const int LEDEyeRightPin = 10;
const int LEDEyeCenterPin = 11;
const int LEDEyeLeftPin = 12;
const long glanceMillis = 500;

int MicLeftVal = 0;
int MicCenterVal = 0;
int MicRightVal = 0;
int MicDirectionVal = 0;
int MicDirectionIndex = 0;
int PIRLeftVal = 0;
int PIRRightVal = 0;
long lastMillis = 0;
unsigned long currentMillis = millis();

void setup() {
```

```
    NeckServo.attach(NeckServoPin);
    pinMode(PIRRightPin,INPUT);
    pinMode(PIRLeftPin,INPUT);
    pinMode(LEDEyeRightPin,OUTPUT);
    pinMode(LEDEyeCenterPin,OUTPUT);
    pinMode(LEDEyeLeftPin,OUTPUT);
    NeckCenter();
    EyesCenter();
}

void loop(){
  MicLeftVal = analogRead(MicLeftPin);
  MicCenterVal = analogRead(MicCenterPin);
  MicRightVal = analogRead(MicRightPin);
  PIRLeftVal = digitalRead(PIRLeftPin);
  PIRRightVal = digitalRead(PIRRightPin);

  currentMillis = millis();

  if(MicLeftVal >= MicTrigger || MicRightVal >= MicTrigger){
     MicDirectionVal = MicRightVal - MicLeftVal;
  if(MicDirectionVal >= 0)
     MicDirectionIndex = IndexSteps + ((MicDirectionVal *
     IndexSteps) / MicRightVal);
  else
     MicDirectionIndex = IndexSteps + ((MicDirectionVal *
     IndexSteps) / MicLeftVal);

if (currentMillis - lastMillis > glanceMillis){
reset();
if (MicDirectionIndex >=4)
EyesRight();
else if  (MicDirectionIndex <=2)
EyesLeft();
}
 }
else
if (currentMillis - lastMillis > glanceMillis){
  if  (MicCenterVal >= MicTrigger)
  EyesFlash();
  else
  EyesCenter();
```

```
}
}

/*

----------------------------------------------------------
 NeckCenter()
 NeckRight()
 NeckLeft()
 EyesCenter()
 EyesLeft()
 EyesRight()
 EyesFlash()
 reset()
----------------------------------------------------------
*/

void NeckCenter(){
  NeckServo.write(90); }

void NeckRight(){
  NeckServo.write(179); }

void NeckLeft(){
  NeckServo.write(0); }

void EyesCenter() {
digitalWrite(LEDEyeRightPin, LOW);
digitalWrite(LEDEyeCenterPin, HIGH);
digitalWrite(LEDEyeLeftPin, LOW);
NeckCenter();
}

void EyesLeft() {
digitalWrite(LEDEyeRightPin, LOW);
digitalWrite(LEDEyeCenterPin, LOW);
digitalWrite(LEDEyeLeftPin, HIGH);
if (PIRLeftVal==HIGH) NeckLeft();
}

void EyesRight() {
digitalWrite(LEDEyeRightPin, HIGH);
digitalWrite(LEDEyeCenterPin, LOW);
```

```
  digitalWrite(LEDEyeLeftPin, LOW);
  if (PIRRightVal == HIGH) NeckRight();
}

void EyesFlash() {
  digitalWrite(LEDEyeRightPin, LOW);
  digitalWrite(LEDEyeLeftPin, LOW);
  digitalWrite(LEDEyeCenterPin, LOW);
  delay (200);
  digitalWrite(LEDEyeCenterPin, HIGH);
  delay (200);
  digitalWrite(LEDEyeCenterPin, LOW);
  delay (200);
  digitalWrite(LEDEyeCenterPin, HIGH);
}

void reset() {
  lastMillis = currentMillis;
}
```

Starting at the top, the constants and variables for the servo and PIR devices are defined. During setup, the servo is attached, and the pins used for the PIR sensors are defined as inputs. In the main loop, the microphones and PIR sensors are read before zeroing the elapsed-time variable `currentMillis`. The acoustic data are handled as in the original system, comparing amplitudes of the signals from the three microphones to determine the location of the sound source.

The contribution of the PIR data is made in the subroutines, according to the following logic:

PIDR + sound detected on right: Neck rotates toward the right.
PIRR only: No neck rotation.
PIRR + sound detected on left: No neck rotation.

Operation

Use your studio monitor or the "click click" of your pen to trigger the microphones, as before. However, now pay attention to where you stand. From behind the unit, you should be able to trigger the microphones without setting off the PIR sensors. Next, stand on the structure's left (your right if you're facing the foam roller or mannequin), and make some noise. Then repeat the exercise while standing on the structure's right.

For a bonus activity, use the index identified for each sound to define different degrees of rotation. Just remember that the servo and turret have built-in refractory periods because it takes time for the pan to rotate back in place after a PIR sensor has been triggered.

Increased Intensity Discrimination

This modification focuses on the components and infrastructure construction and uses the same code as the preceding example. The goal is to increase the intensity discrimination of the system so that there is, for example, a greater signal-to-noise ratio overall and a larger difference between a microphone facing the source and the one facing away from the source.

Bill of Materials

To maximize the intensity discrimination of your system and thereby increase the accuracy of localization, you'll need the following:

- Silicone pinch bowls (3)
- Speaker crossover circuit
- Directional electret microphone elements (3)

Construction

The most important modification is the substitution of directional electret microphone elements for the standard omnidirectional elements that are probably in your microphone assembly. Unidirectional electret microphone elements, such as the CMI-5247TF-K, are inexpensive and readily available. If you order a dozen from your favorite parts supplier, you'll probably end up paying more for postage than for the parts proper.

The theoretical directional and frequency responses of a typical unidirectional electret microphone element are shown in Figure 5-11. However, *theoretically* means a microphone element suspended by a monofilament thread in an acoustically isolated room. Simply mounting a omnidirectional microphone onto a PC board distorts the directivity pattern.

As shown in Figure 5-11, the directional characteristics are more pronounced at 1 kHz and above. At 1 kHz, the response from the rear of the unidirectional microphone element depicted in the figure is 25 dB below that of the forward response. At 125 Hz, the difference in response is only 10 dB. Recall that the response of electret microphone elements is a function of sound pressure level, and 20 dB is equivalent to a 10-fold difference in level. As such, the 10-dB difference in response is only a threefold difference in sound pressure level.

Figure 5-12 shows a partially modded Grove integrated microphone amplifier module with the original omnidirectional microphone element (*above*) and the unidirectional CMI-5247TF-K electret microphone element (*below*). Note that the unidirectional microphone should be attached so that it points away from the connector and toward the sound source.

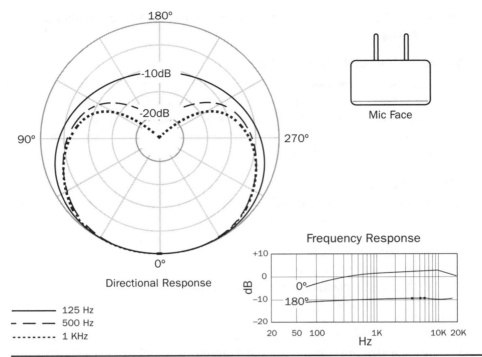

FIGURE 5-11   Theoretical directional and frequency responses of a typical unidirectional electret microphone element.

FIGURE 5-12   Grove integrated microphone amplifier module in midmod.

Although the unidirectional microphone element is a big improvement over an omnidirectional element, you can achieve another 3 to 6 dB of relative isolation by mounting the integrated assembly in a small silicone pinch bowl. These soft, flexible bowls, used to hold salt, pepper, and other spices, are about the same diameter and consistency as a human ear. Mount an amplifier assembly on the bottom of each of the three bowls, and solder the microphone leads to the unidirectional element that's mounted at the focal point in the bowl. Then fold one lip over to create an external humanoid ear. See Figure 5-13 for details.

The last part of this modification involves adding a high-pass filter between the microphone amplifier output and the Arduino's analog input pins. Because the low-frequency sounds contribute relatively little to localization, they are essentially noise. As we saw in Figure 5-11, the benefit of a unidirectional microphone is limited to about 1 kHz and above.

There are several avenues for adding filtering to the microphone signal. One is to change components in the microphone amplifier so that lower-frequency signals are amplified less than high-frequency signals. Another option is to use a multichannel graphic equalizer, commonly found in home and professional recording studios. Graphic equalizers sold as guitar foot pedals are often less expensive than desktop units. An electronic digital audio workstation (DAW)

FIGURE 5-13 Unidirectional microphone amplifier module mounted in a silicone pinch bowl with lip rolled over to form a directional ear.

software program and interface to your computer constitute another option. We've also had good results with repurposed two- or three-way speaker crossovers. The two channels are used to filter signals from the left and right microphones.

The advantage of a speaker crossover is that there are no batteries and little added noise. Crossovers also readily available online. The primary downside of a crossover is bulk. A typical crossover is at least the size of an Arduino Uno. A secondary issue is impedance matching. Most crossover circuits are designed for low-impedance input and output. If your microphone amplifier is intolerant of low output impedance, then you'll either have to modify the circuit or use an impedance transformer.

Operation

The system should operate just as before—only better. Because of the enhancements, there should be fewer false triggers and more accurate localization. The directivity afforded by the unidirectional microphone elements and the silicone cup "ears" should allow you to increase the degree of indexing in the amplitude-comparison algorithm and thereby move the head to intermediate positions instead of only left or right. Figure 5-14 shows the platform ready for action. Note the discrete LED "eyes" next to the pan base.

Psychoacoustics

This software-only modification of the basic amplitude-based localization program enables you to explore the psychoacoustic phenomenon of the duration-amplitude linkage described earlier. In addition to the main trigger level, there's a lower trigger level that will fire when the duration of the sound exceeds the value you define.

Code

The Arduino code that supports duration-linked triggering is shown in Listing 5-3.

LISTING 5-3   Arduino code for amplitude-based sound localization with psychoacoustic modification for duration.

```
/*
Amplitude-based sound localization with psychoacoustics
modification
Androids: Build Your Own Lifelike Robots by Bergeron and Talbot
Listing 5-3. See www.mhprofessional.com/Androids for fully
documented code
Arduino 1.0.1 environment
*/

#include <Servo.h>
```

FIGURE 5-14 Amplitude-based localization platform with PIR sensors, servo-controlled pan, and unidirectional microphones mounted in silicone "ears." The pan platform is rotated fully counterclockwise.

```
Servo NeckServo;

const int MicLeftPin = A0;
const int MicCenterPin = A1;
const int MicRightPin = A2;
const int MicTriggerLow = 300;
const int MicTriggerHigh = 500;
const int IndexSteps = 3;
```

```
const int PIRRightPin = 5;
const int PIRLeftPin = 6;
const int NeckServoPin = 9;
const int LEDEyeRightPin = 10;
const int LEDEyeCenterPin = 11;
const int LEDEyeLeftPin = 12;
const long glanceMillis = 500;
const long LowSoundMillis = 300;
const int LowSoundDelay = 300;

int MicLeftVal = 0;
int MicCenterVal = 0;
int MicRightVal = 0;
int MicDirectionVal = 0;
int MicDirectionIndex = 0;
int PIRLeftVal = 0;
int PIRRightVal = 0;
long lastMillis = 0;
long durationMillis = 0;
unsigned long currentMillis = millis();

void setup() {
  NeckServo.attach(NeckServoPin);
  pinMode(PIRRightPin,INPUT);
  pinMode(PIRLeftPin,INPUT);
  pinMode(LEDEyeRightPin,OUTPUT);
  pinMode(LEDEyeCenterPin,OUTPUT);
  pinMode(LEDEyeLeftPin,OUTPUT);
  NeckCenter();
  EyesCenter();
  resetDuration();
}

void loop(){
MicLeftVal = analogRead(MicLeftPin);
MicCenterVal = analogRead(MicCenterPin);
MicRightVal = analogRead(MicRightPin);

/*
For testing duration volume
MicLeftVal = min (MicLeftVal, 350);
MicCenterVal = min (MicCenterVal, 350);
```

```
   MicRightVal = min (MicRightVal, 350);
*/

   PIRLeftVal = digitalRead(PIRLeftPin);
   PIRRightVal = digitalRead(PIRRightPin);

   currentMillis = millis();

   if (MicLeftVal >= MicTriggerHigh || MicRightVal >=
       MicTriggerHigh){
       MicDirectionVal = MicRightVal - MicLeftVal;
   if(MicDirectionVal >= 0)
       MicDirectionIndex = IndexSteps + ((MicDirectionVal *
       IndexSteps) / MicRightVal);
   else
       MicDirectionIndex = IndexSteps + ((MicDirectionVal *
       IndexSteps) / MicLeftVal);

if (currentMillis - lastMillis > glanceMillis){
reset();
if (MicDirectionIndex >=4)
EyesRight();
else if  (MicDirectionIndex <=2)
EyesLeft();
else
EyesFlash();
}
 }

else if (MicLeftVal >= MicTriggerLow || MicRightVal >=
         MicTriggerLow){
   delay (LowSoundDelay);
   MicLeftVal = analogRead(MicLeftPin);
   MicCenterVal = analogRead(MicCenterPin);
   MicRightVal = analogRead(MicRightPin);
   MicLeftVal = min (MicLeftVal, 350);
   MicCenterVal = min (MicCenterVal, 350);
   MicRightVal = min (MicRightVal, 350);

   if (MicLeftVal >= MicTriggerLow || MicRightVal >= MicTriggerLow){
MicDirectionVal = MicRightVal - MicLeftVal;
   if(MicDirectionVal >= 0)
```

```
      MicDirectionIndex = IndexSteps + ((MicDirectionVal *
      IndexSteps) / MicRightVal);
    else
      MicDirectionIndex = IndexSteps + ((MicDirectionVal *
      IndexSteps) / MicLeftVal);

  if (currentMillis - lastMillis > glanceMillis - LowSoundMillis){
  reset();
  if (MicDirectionIndex >=4)
  EyesRight();
  else if  (MicDirectionIndex <=2)
  EyesLeft();
  else
  EyesFlash();
  }
    }
  }
  else
  if (currentMillis - lastMillis > glanceMillis - LowSoundMillis){
    if  (MicCenterVal >= MicTriggerLow)
    EyesFlash();
    else
    EyesCenter();
  }
  }

  /*
    ----------------------------------------------------------
    NeckCenter()
    NeckRight()
    NeckLeft()
    EyesCenter()
    EyesLeft()
    EyesRight()
    EyesFlash()
    reset()
    resetDuration()
    ----------------------------------------------------------
    */

  void NeckCenter(){
    NeckServo.write(90);
```

```
}

void NeckRight(){
  NeckServo.write(179);
}

void NeckLeft(){
  NeckServo.write(0);
  }


void EyesCenter() {
digitalWrite(LEDEyeRightPin, LOW);
digitalWrite(LEDEyeCenterPin, HIGH);
digitalWrite(LEDEyeLeftPin, LOW);
NeckCenter();
}

void EyesLeft() {
digitalWrite(LEDEyeRightPin, LOW);
digitalWrite(LEDEyeCenterPin, LOW);
digitalWrite(LEDEyeLeftPin, HIGH);
if (PIRLeftVal==HIGH) NeckLeft();
delay (300);
}

void EyesRight() {
digitalWrite(LEDEyeRightPin, HIGH);
digitalWrite(LEDEyeCenterPin, LOW);
digitalWrite(LEDEyeLeftPin, LOW);
if (PIRRightVal == HIGH) NeckRight();
delay (300);
}

void EyesFlash() {
  digitalWrite(LEDEyeRightPin, LOW);
  digitalWrite(LEDEyeLeftPin, LOW);
  digitalWrite(LEDEyeCenterPin, LOW);
  delay (200);
  digitalWrite(LEDEyeCenterPin, HIGH);
  delay (200);
  digitalWrite(LEDEyeCenterPin, LOW);
```

```
    delay (200);
    digitalWrite(LEDEyeCenterPin, HIGH);
    NeckCenter();
    delay (300);
}

void reset() {
    lastMillis = currentMillis;
}

void resetDuration() {
    durationMillis = currentMillis;
}
```

The key additions to the basic amplitude-based localization program include the constants `MicTriggerLow` and `MicTriggerHigh`, which define the duration-linked trigger level and the duration-independent trigger level, respectively. When the detected sound amplitude equals or exceeds the level defined by `MicTriggerHigh`, the system is triggered, regardless of the sound duration. However, if the sound level is greater than `MicTriggerLow` and less than `MicTriggerHigh`, then the sound duration must equal or exceed the time defined in the constant `LowSoundMillis`.

The logic for this addition is defined in the second `else if` routine within the main loop. As you can see, with the exception noted earlier, the code section is essentially identical to the duration-independent logic. We opted not to condense the code in the name of transparency and to provide you with a framework in which you can experiment with different `IndexSteps` values and glance durations.

Note the commented section in the main loop that can be used for testing the duration-dependent code. It limits the value returned by `analogRead` for each of the microphone connections, ensuring a level less than defined by `MicTriggerHigh`.

Operation

In operation, you should be able to use lower-level, longer-duration tones to trigger the system. The relative durations and sound level are up to you. Once you get a feel for the interaction of duration and level, try to emulate typical human responses. For example, try using 50- and 500-ms tones of different intensities. This is where an SPL meter would come in handy. Otherwise, you can use the `print` function to see what section of code is being triggered.

Time-Based Localization

Now that you're comfortable with amplitude-based localization, we're going to move to time-based localization. Actually, it's more of a hybrid approach that depends primarily on the relative arrival times of sounds at the microphones but requires sounds of amplitude great enough to be detected by the integrated microphone amplifiers.

Because the A/D conversion process performed by the analog ports on the 16-MHz Arduino Uno is slow compared with the time intervals involved in time-based localization, we'll use the much faster digital I/O. This requires the addition of a pair of optoisolators that essentially perform instantaneous 1-bit A/D conversion and modest changes to the amplitude-based code.

Bill of Materials

Assuming that you're starting from scorched earth, you'll need the following to enter the realm of time-based localization:

- Electret microphones with amplifiers (2)
- Discrete LEDs (6)
- 4N35 optoisolators (2)
- 22-nF (0.022-µF) @ 10-V ceramic or polyester film capacitors (2)
- Arduino Uno
- 500-Ω, ¼-W resistor
- 100-Ω, ¼-W resistors (2)
- 5-Vdc power supply
- Jumpers or wires
- Acoustic signal source
- Head-sized support structure
- Acoustically dead environment
- Breadboard or prototyping shield
- SPL meter (optional)

Figure 5-15 shows the major additional components of the experimental setup, including the 4N35 optoisolator, which is an encapsulated gallium-arsenide IR diode that's optically coupled to a silicon phototransistor. Typical turn-on time is about 8 ms—more than adequate for our purposes. Feel free to substitute other optoisolators for the 4N35, including multichannel integrated circuits (ICs) that may be less expensive and require less breadboard real estate. The capacitor type isn't critical. What's important is that the capacitors are of the same type, value, and rating.

Circuit

The circuit for the first experiment with time-based localization, shown in Figure 5-16, consists of a standard Arduino Uno microcontroller, our six-LED "eye" assembly, and only two connections to the digital ports of the Arduino. When a microphone module picks up a sound of significant amplitude, the resulting signal

FIGURE 5-15   Major additional components of the authors' experimental setup.

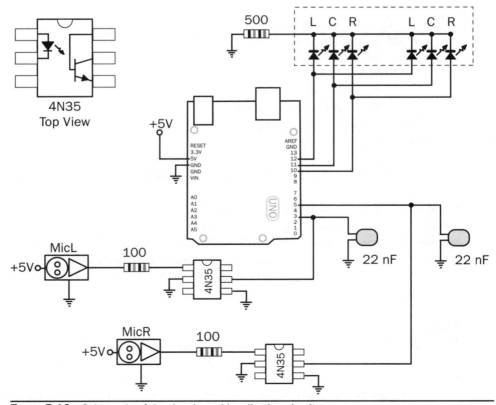

FIGURE 5-16   Schematic of the time-based localization circuit.

is sent to the IR LED pin of the corresponding 4N35. With sufficient drive, the LED illuminates a silicon phototransistor thtat drives the associated digital port low.

The 22-nF capacitor reduces noise and false triggering of the port. It also establishes a time constant for the decay of voltage when the 4N35's phototransistor is triggered, pulling the digital input low. Feel free to experiment with other capacitor values.

Construction

This is a perfect project for breadboard. Both optoisolator circuits fit nicely onto one of the minibreadboards with adhesive backing sold for the Arduino prototyping shields. Once you're happy with capacitor values for your particular setup, you might want to solder the components directly to the prototyping shield. However, wait until you've finished the other modifications in this chapter—you'll need to repurpose those 4N35s.

Code

The code for this first iteration of time-based localization parallels that of the stripped-down version of the code first introduced for amplitude-based localization. In addition, because of the computational overhead and to keep the code simple, we're limiting the system to two microphone modules.

LISTING **5-4**   Arduino code for time-based localization.

```
/*
Time-Based Localization
Androids: Build Your Own Lifelike Robots by Bergeron and Talbot
Listing 5-4. See www.mhprofessional.com/Androids for fully
documented code
Arduino 1.0.1 environment
*/

const int MicLeftPin = 3;
const int MicRightPin = 5;
const int LEDEyeRightPin = 10;
const int LEDEyeCenterPin = 11;
const int LEDEyeLeftPin = 12;

unsigned long currentMillis = millis();
int MicLState = 0;
int MicRState = 0;

void setup() {
  pinMode(LEDEyeRightPin,OUTPUT);
```

```
     pinMode(LEDEyeCenterPin,OUTPUT);
     pinMode(LEDEyeLeftPin,OUTPUT);
     pinMode(MicLeftPin, INPUT);
     pinMode(MicRightPin, INPUT);
     digitalWrite(MicLeftPin,HIGH);
     digitalWrite(MicRightPin,HIGH);
}

void loop(){
MicLState = 0;
MicRState = 0;
MicRState = !digitalRead(MicRightPin);
MicLState = !digitalRead(MicLeftPin);
if (MicRState || MicLState) ProcessTime();
}

/*
  --------------------------------------------------------
  ProcessTime()
  EyesCenter()
  EyesLeft()
  EyesRight()
  EyesFlash()
  --------------------------------------------------------
  */

void ProcessTime() {
if (MicRState && MicLState) {
EyesFlash();
}
else
if (MicLState){
EyesLeft();
}
else
if (MicRState){
EyesRight();
}
else {
EyesCenter();
}
}
```

```
void EyesCenter() {
digitalWrite(LEDEyeRightPin, LOW);
digitalWrite(LEDEyeCenterPin, HIGH);
digitalWrite(LEDEyeLeftPin, LOW);
}

void EyesLeft() {
digitalWrite(LEDEyeRightPin, LOW);
digitalWrite(LEDEyeCenterPin, LOW);
digitalWrite(LEDEyeLeftPin, HIGH);
delay (500);
EyesCenter();
}

void EyesRight() {
digitalWrite(LEDEyeRightPin, HIGH);
digitalWrite(LEDEyeCenterPin, LOW);
digitalWrite(LEDEyeLeftPin, LOW);
delay (500);
EyesCenter();
}

void EyesFlash() {
  digitalWrite(LEDEyeRightPin, LOW);
  digitalWrite(LEDEyeLeftPin, LOW);
  digitalWrite(LEDEyeCenterPin, LOW);
  delay (200);
  digitalWrite(LEDEyeCenterPin, HIGH);
  delay (200);
  digitalWrite(LEDEyeCenterPin, LOW);
  delay (200);
  digitalWrite(LEDEyeCenterPin, HIGH);
  delay (300);
}
```

The key variables defined at the top of the listing are the microphone state variables MicLState and MicRState. They indicated whether the microphone modules have been triggered. Notice also that the pins assigned to the microphone modules, MicLeftPin and MicRightPin, are defined as input pins and are pulled high with the internal resistors using the digitalWrite function. If you refer to the circuit diagram, you can see that when either microphone module triggers an optoisolator, the phototransistor conducts, bringing the Arduino's digital input pin low. When it's quiet, the input pins are high.

Given the time-critical nature of the program, the main loop is short and simple. The microphone state variables are set to zero, and the corresponding digital input pins are polled. If either of the pins changes state, the loop is exited to execute the `ProcessTime()` subroutine. This subroutine parallels the code used in Listing 5-1, with calls to the eye-movement subroutines to indicate whether the sound source is to the right, left, or center. However, instead of polling a third, centrally located microphone module, the code checks for simultaneous changes in `MicLState` and `MicRState` to indicate that the source is to the center.

Operation

As with the amplitude-based experiments, you can go low tech or high tech. A retractable pen that makes a nice, loud "click click" makes a great high-frequency sound source. Click nearer to the right microphone module, and the eyes should indicate right, and vice versa.

You'll probably notice a decrease in accuracy relative to the amplitude-based localization systems developed earlier. This is so because even though we're working with digital input, the time required for a read and the time between reads are significant compared with the signals we're trying to detect. In other words, the system is still too slow. Fortunately, we have the technology to fix it.

Increased Temporal Resolution

The standard 16-MHz Arduino Uno is no speed demon, and that's okay for most applications. However, we need every clock cycle to count if we're going to achieve accuracies approaching what we had with the amplitude-based localization projects.

Fortunately, we have several options. The first is to replace the slow, sequential read of the digital input pins with simultaneous parallel reads of the input pins using direct port manipulation. Because of limitations in how the ports are read directly, we'll have to rewire the optoisolator circuits to provide a low input when it's quiet and drive the input pin high when the optoisolator is triggered.

Bill of Materials

Assuming that you've built the preceding system, the only additional hardware you need is a pair of 10-kΩ, ¼-W resistors.

Circuit

The circuit changes from the basic time-based localization circuit are shown in Figure 5-17. The 10-kΩ resistors are used to pull the digital input pins low when the optoisolators are inactive. When a phototransistor in an optoisolator is triggered, the digital pin is brought high. As with the preceding circuit, you can experiment

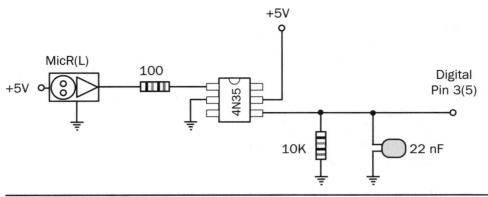

Figure 5-17   Circuit changes to optoisolator output.

with different capacitor values, but 22 nF is a good compromise between activation time and smoothing the audio-frequency fluctuations to a reasonable dc signal.

Construction
Construction involves simply rearranging the optoisolator circuit and adding the 10-kΩ resistors. Again, this is a perfect application for a breadboard.

Code
The Arduino code for time-based localization using direct port manipulation of the I/O ports is shown in Listing 5-5.

Listing 5-5   Arduino code for time-based localization using direct port manipulation.

```
/*
Time-Based Localization Using Direct Port Manipulation
Androids: Build Your Own Lifelike Robots by Bergeron and Talbot
Listing 5-5. See www.mhprofessional.com/Androids for fully
documented code
Arduino 1.0.1 environment
*/

const int MicLeftPin = 3;
const int MicRightPin = 5;
const int a = 40;
const int b = 8;
const int c = 32;
const int LEDEyeRightPin = 10;
const int LEDEyeCenterPin = 11;
const int LEDEyeLeftPin = 12;
```

```
unsigned long currentMillis = millis();
int MicLState = 0;
int MicRState = 0;
int z = 0;

void setup() {
  pinMode(LEDEyeRightPin,OUTPUT);
  pinMode(LEDEyeCenterPin,OUTPUT);
  pinMode(LEDEyeLeftPin,OUTPUT);

  pinMode(MicLeftPin, INPUT);
  pinMode(MicRightPin, INPUT);
}

void loop(){
  z = a & PIND;
  if (z) ProcessTime();
 }

/*
  --------------------------------------------------------
  ProcessTime()
  EyesCenter()
  EyesLeft()
  EyesRight()
  EyesFlash()
  --------------------------------------------------------
  */

void ProcessTime() {
if (z == a) {
EyesFlash();
}
else
if (z==b){
EyesLeft();
}
else
if (z==c){
EyesRight();
```

```
}
else {
EyesCenter();
}
}

void EyesCenter() {
digitalWrite(LEDEyeRightPin, LOW);
digitalWrite(LEDEyeCenterPin, HIGH);
digitalWrite(LEDEyeLeftPin, LOW);
}

void EyesLeft() {
digitalWrite(LEDEyeRightPin, LOW);
digitalWrite(LEDEyeCenterPin, LOW);
digitalWrite(LEDEyeLeftPin, HIGH);
delay (500);
EyesCenter();
}

void EyesRight() {
digitalWrite(LEDEyeRightPin, HIGH);
digitalWrite(LEDEyeCenterPin, LOW);
digitalWrite(LEDEyeLeftPin, LOW);
delay (500);
EyesCenter();
}

void EyesFlash() {
  digitalWrite(LEDEyeRightPin, LOW);
  digitalWrite(LEDEyeLeftPin, LOW);
  digitalWrite(LEDEyeCenterPin, LOW);
  delay (200);
  digitalWrite(LEDEyeCenterPin, HIGH);
  delay (200);
  digitalWrite(LEDEyeCenterPin, LOW);
  delay (200);
  digitalWrite(LEDEyeCenterPin, HIGH);
  delay (300);
}
```

This listing illustrates the reason many people avoid direct port access—undocumented code can be difficult to follow, and it's relatively fragile. Change the pin assignments, and everything breaks.

As with cooking, most of the work here is in the preparation. There's not much to the main loop, as you can see. The first five lines, however, are critical. The values assigned to constants a, b, and c are based on the pin assignments for `MicLeftPin` and `MicRightPin`.

We are using port D (`PIND` in the main loop), which includes pins 0–7. The main loop simply checks to see if either pin 3 or pin 5 is `HIGH`. What's potentially confusing is that this is performed in binary.

Assuming that only pin 3 of port D is `HIGH`, the state of pins 0–7 looks like this:

```
0 0 0 0 1 0 0 0
```

That is, from right to left, pin 0 has a value of 0, pin 1 has a value of 0, pin 2 has a value of 0, and pin 3 has a value of 1. It happens that the decimal equivalent of B00001000 is $2^3 = 8$, which is why the constant b is assigned the value 8.

Similarly, if only pin 5 of port D is `HIGH`, the state of pins 0–7 looks like this:

```
0 0 1 0 0 0 0 0
```

Remember, pin 0 is at the far right, and pin 7 is at the far left. B00100000 is equivalent to $2^5 = 32$, the value assigned variable c.

Now, if we want to determine whether pin 3 or 5 is `HIGH`, we can perform a Boolean AND operation on pins 3 and 5. Because we can perform this in a single operation by comparing the pin values with B00100000 + B00001000 = B00101000 $= 2^3 + 2^5 = 8 + 32 = 40$. And 40 is assigned to variable a. Now, if we perform a Boolean AND on the current port state and the Boolean equivalent of 40, we can detect a `HIGH` state.

For example, let's say that pin 3 brought `HIGH` by activity from the right microphone module. We have the following comparison:

```
a        B00101000
PIND     B00001000
&
z        B00001000
```

So, with pin 3 `HIGH` (B0001000), the Boolean AND with a (B00101000), we get a nonzero result for z. If we had used pins 3 (B00001000) and 4 (B00010000), then we would have assigned variable a the value of B00011000 $= 2^3 + 2^4 = 8 + 16 = 24$.

Operation

Whether you click your Bic or use a speaker to play the MP3 files provided online, you should see a significant improvement in accuracy with direct port access.

Benchmarks

At this point, you may be asking whether it's worth the bother of working with direct port manipulation, especially if you're not used to working with Boolean operations. To help you decide, we've put together a little benchmarking program in Listing 5-6 that compares analog, digital, and direct port read times.

LISTING 5-6   Arduino code for benchmarking relative analog, digital, and direct port read times.

```
/*
Analog, Digital and Direct Port Read Times
Androids: Build Your Own Lifelike Robots by Bergeron and Talbot
Listing 5-6. See www.mhprofessional.com/Androids for fully
documented code
Arduino 1.0.1 environment
*/

const int MicLeftPin = 3;
const int MicRightPin = 5;
const int a = 40;
const int b = 8;
const int c = 32;

unsigned long currentMillis = millis();
int MicLState = 0;
int MicRState = 0;
int z = 0;

void setup() {
pinMode(MicLeftPin, INPUT);
pinMode(MicRightPin, INPUT);
Serial.begin(9600);
}

void loop(){
Serial.print("Analog Read: ");
currentMillis = millis();
for (int i = 0; i <10000; i++){
MicRState = analogRead(MicRightPin);
```

```
MicLState = analogRead(MicLeftPin);
   }
Serial.println(millis() - currentMillis);
delay (100);
Serial.print("Digital Read: ");
currentMillis = millis();
for (int i = 0; i <10000; i++){
MicRState = !digitalRead(MicRightPin);
MicLState = !digitalRead(MicLeftPin);
   }
  Serial.println(millis() - currentMillis);
  delay (100);
  Serial.print("Direct Port Read: ");
currentMillis = millis();
for (int i = 0; i <10000; i++){
z = a & PIND;
}
Serial.println(millis() - currentMillis);
}
```

The benchmarking code runs through 10,000 iterations for each of the three methods we discussed for reading ports on the Arduino. Keep in mind that the benchmark code is a stripped-down version of what's used in the full sound-localization programs and, as such, represents the best performance. With a fully configured program, expect somewhat less than best performance. We get the following results:

Analog read: 2,240
Digital read: 108
Direct port read: 7

Based on our benchmark results, there is a 2,240/10,000 or 224-ms interval between analog port reads, 108/10,000 or 10.8-ms interval between digital port reads, and 7/10,000 or 0.7-ms interval between direct port reads. That is, the maximum sample rate that we can achieve with the Uno is 1/0.7 ms or 143 Hz. This compares to 1/224 ms or 4.5 Hz for the analog read.

We can also convert the intervals into feet. Using the 343.2 m/s (1,126 ft/s) figures discussed earlier and converting time to distance covered by sound, we get 76 m (252 ft), 3.7 m (12 ft), and 0.2 m (0.7 ft) for analog, digital, and direct port reads, respectively. Clearly, the chances that our localization system will detect the front of a sound pressure wave from the click of a pen striking one ear before the next are slim for the standard analog and digital port reads. On the other hand, 0.7 ft or about 8 in. isn't that bad. As you'll see working with the experimental setup, the direct port read method produces consistent results. How can this be, given the relatively poor temporal resolution?

Remember that this is a hybrid system that responds to both amplitude and first-to-arrive timing. Even if the system misses the wave front of a sound, the amplitude will still be greatest toward the source. And, all else being equal, the microphone module with the greatest signal strength will trigger the optoisolator first.

Listing 5-6 is a good place to begin for experimentation with execution time. For example, what's the cost, in time, for switching the logic on the digital reads (`!digitalRead`) compared with straight digital reads? Would a different Boolean operator provide a speed advantage in the direct port reads? Can you devise a fourth software-only method to minimize time to completion?

Heavy Iron

One of the advantages of the Arduino platform is that there is an abundance of clones on the market. Some are smaller, some have more peripherals, and at least one, the 80-MHz ChipKit UNO32, is a speed demon. The inexpensive Arduino-compatible processor board, shown next to a standard Uno in Figure 5-18, allows us to throw hardware at the speed problem.

FIGURE 5-18 80-MHz ChipKit UNO3 (*left*) and 16-MHz Arduino Uno (*right*).

Unfortunately, software compatibility is limited. At the time of this writing, the ChipKit IDE MPIDE 0023 isn't fully compatible with the Arduino 1.0.1 IDE. As a result, some of the latest libraries for the Arduino may not work. Still, the MPIDE IDE looks and feels like the older Arduino IDE, and unless you're doing something odd, then the MPIDE IDE and resulting code should be fine.

LISTING **5-7**   Code for the 80-MHz ChipKit UNO32 comparing time-based localization using standard analog and digital read statements.

```
/*
Time-Based Localization; Speed Evaluation of ChipKit UNO32
Androids: Build Your Own Lifelike Robots by Bergeron and Talbot
Listing 5-7. See www.mhprofessional.com/Androids for fully
documented code
MPIDE 0023 environment
*/

const int MicLeftPin = 3;
const int MicRightPin = 5;

unsigned long currentMillis = millis();
int MicLState = 0;
int MicRState = 0;

void setup() {
pinMode(MicLeftPin, INPUT);
pinMode(MicRightPin, INPUT);
Serial.begin(9600);
}

void loop(){
Serial.print("Analog Read: ");
currentMillis = millis();
for (int i = 0; i <10000; i++){
MicRState = analogRead(MicRightPin);
MicLState = analogRead(MicLeftPin);
}
Serial.println(millis() - currentMillis);
delay (100);

Serial.print("Digital Read: ");
currentMillis = millis();
for (int i = 0; i <10000; i++){
```

```
MicRState = !digitalRead(MicRightPin);
MicLState = !digitalRead(MicLeftPin);
}
Serial.println(millis() - currentMillis);
}
```

The code in Listing 5-7 parallels the benchmark program provided earlier for the generic Arduino. On our run of the benchmark, we had the following results:

Analog read: 100
Digital read: 10

This translates to a 100/10,000 or 10-ms interval between analog port reads and a 10/10,000 or 1-ms interval between digital port reads. Continuing with our assumption that sound travels at 343.2 m/s (1,126 ft/s) in air at 20°C (68°F), the distance traveled by sound between analog reads is 3.43 m or a little over 11 ft. This compares with 0.34 m or a little over 1 ft between digital reads. Ideally, we'd have an interval of only a few centimeters, but, as with the direct port read with the Arduino Uno, the standard digital read on the UNO32 isn't too bad. It approaches the direct port read interval we achieved with the Arduino Uno.

Although the UNO32 is definitely faster than the Arduino Uno, the ChipKit application still uses sequential, as opposed to parallel, reads for both analog and digital inputs. And unfortunately, reading the ports directly falls under the category of "odd" for programming the UNO32. Although the ChipKit UNO32 fully supports direct port access, and the resulting performance is superb, the code is hardware-specific and therefore incompatible with the current Arduino Uno and Mega. What you develop at the hardware level for the UNO32 stays with the UNO32.

If you're considering a hardware upgrade to the basic Arduino, you should also explore the Maple from LeafLabs. This Arduino-compatible hardware is built around a 72-MHz ARM chip, with 29 digital and 16 analog I/O pins. The open-source IDE is similar to the Arduino IDE but not directly compatible. As with the ChipKit boards, if you want to read and write to the hardware ports, your code won't port directly to an Arduino.

Eye Variations

Up to this point, we've used pupil location and neck rotation as means of indicating the results of the sound localization. For clarity, we've left out the PIR detectors and neck rotation from the time-based localization examples, but they're directly applicable. Another addition that's applicable to both amplitude- and time-based localization is improved aesthetics.

At some point in your android design, you'll probably substitute movable cameras in the eye sockets for the simple trio of discrete LEDs used to represent

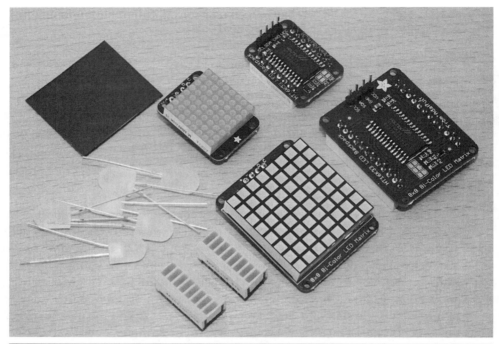

Figure 5-19   Examples of LED-based options for localization feedback.

pupil position. As an intermediate step, you can provide a more realistic representation of eye movement by using other forms of LEDs. For example, as shown in Figure 5-19, we've tried large, high-intensity discrete LEDs, LED bars, as well as monochrome and color LED matrices. We've also worked with small organic LED (OLED) displays. For now, let's look at the monochrome and color LED matrices, which provide a great look for little cost or computational overhead.

Bill of Materials

Let's start with an Adafruit mini 8 × 8 LED matrix and then explore the larger Adafruit bicolor 8 × 8 LED matrix. The mini 8 × 8 fits nicely into the eye sockets of a life-sized mannequin and is available in yellow, blue, red, and green. The Adafruit bicolor 8 × 8 LED matrix has the obvious benefit of supporting multiple simultaneous colors with no additional hardware overhead.

Circuit

The Adafruit LED matrices have two things going for them. First, as illustrated in Figure 5-20, is that both the mini and bicolor 8 × 8 LED matrix are I2C devices. As such, we need only assign two pins to either LED matrix on the Arduino—the default serial data (SDA) and clock (SCL) ports A4 and A5. The backpack processor on each display handles the I2C communications, including the pullup resistors.

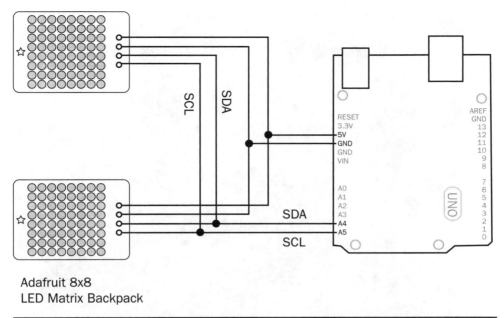

Adafruit 8x8
LED Matrix Backpack

Figure 5-20   Circuit for "eyes" using either the I2C-based Adafruit mini 8 × 8 or the bicolor 8 × 8 LED matrix.

We've tested both I2C displays on the amplitude- and time-based localization programs discussed previously for the Arduino Uno. The ChipKit UNO32 also supports the I2C protocol, but we haven't verified the port addresses.

Construction

There isn't much to construction other than figuring out where to mount the new set of "eyes." Figure 5-21 shows two mini 8 × 8s connected in parallel to an Arduino Uno running on a perf board. Because each mini 8 × 8 has the same I2C address, they can be wired in parallel without modifying the code. The only real limitation is the current supply available through the Arduino, which can easily supply a pair of bicolor or mini 8 × 8 LED units.

The larger bicolor 8 × 8 is 1.2 in. square, a bit larger than needed to fill the cavity of a life-sized eye socket, but it has the advantage of being able to display multiple simultaneous colors. It uses the same wiring as the mini 8 × 8, but you'll have to make minor modifications in the code.

Code

Source code for the mini 8 × 8 and the bicolor 8 × 8 LED matrices are provided in Listings 5-8 and 5-9, respectively. You'll need to add the library files listed below, available from the Adafruit website, to your library folder. And note the matrix address added to the setup routine.

FIGURE 5-21 "Eyes" built with a pair of Adafruit mini 8 × 8 LED matrices. Crosshairs indicate pupil positions.

The main loop is for demonstration purposes only. Use the subroutines in place of the discrete LED subroutines in any of the preceding examples. If you're using a PID-controlled servo pan to rotate the neck, then add the `NeckRight`, `NeckCenter`, and `NeckLeft` calls to the subroutines.

LISTING 5-8 Arduino code for the Adafruit mini 8 × 8 matrix with Adafruit LED matrix backpack.

```
/*
Eyes using the Adafruit Mini 8x8 LED Matrix with Adafruit LED
Matrix backpack
Androids: Build Your Own Lifelike Robots by Bergeron/Talbot
Listing 5-8. See www.mhprofessional.com/Androids for fully
documented code
Arduino 1.0.1 environment
Adafruit Libraries written by Limor Fried/Ladyada for Adafruit
Industries.
*/

#include <Wire.h>
#include "Adafruit_LEDBackpack.h"
```

```
#include "Adafruit_GFX.h"
Adafruit_8x8matrix matrix = Adafruit_8x8matrix();

void setup() {
  matrix.begin(0x70);
}

void loop() {
EyesCenter();
delay (500);
EyesRight();
delay (500);
EyesLeft();
delay (500);
EyesFlash();
delay(500);
}

  /*
  --------------------------------------------------------
  EyesCenter()
  EyesLeft()
  EyesRight()
  EyesFlash()
  --------------------------------------------------------
  */

void EyesCenter() {
  matrix.clear();
  matrix.fillCircle(4,3,1, LED_ON);
  matrix.writeDisplay();
}

void EyesLeft() {
  matrix.clear();
  matrix.fillCircle(6,3,1, LED_ON);
  matrix.writeDisplay();
delay (500);
EyesCenter();
}

void EyesRight() {
```

```
      matrix.clear();
      matrix.fillCircle(2,3,1, LED_ON);
      matrix.writeDisplay();
  delay (500);
  EyesCenter();
  }

  void EyesFlash() {
    matrix.clear();
    matrix.fillCircle(3,3,3, LED_ON);
    matrix.writeDisplay();
    delay (500);
    matrix.clear();
    matrix.fillCircle(4,3,1, LED_ON);
    matrix.writeDisplay();
    delay (300);
  }
```

Calls to the LED matrix are of the form `x,y,r`, where `x` and `y` are the top-left corner of a circle and `r` is the radius. If you prefer black pupils, you can call `drawCircle` instead of `fillCircle`. Furthermore, as you can see in Listing 5-9 for the bicolor 8 × 8 matrix, the only substantive change from the listing for the mini 8 × 8 matrix is the use of `LEDRed`, `LEDGreen`, and `LEDYellow` instead of `LED_ON`.

Both LED matrices are easy to use because of the libraries made freely available by Adafruit. See the source-code listings online for full attributions as well as links to products. And don't forget the option to implement a graphical interface in processing if you're short on time, money, or LEDs.

Listing 5-9 Arduino code for Adafruit bicolor 8 × 8 matrix.

```
/* Eyes using the Adafruit Bicolor 8x8 Matrix with Adafruit LED
Matrix backpack
Androids: Build Your Own Lifelike Robots by Bergeron/Talbot
Listing 5-9.
Arduino 1.0.1 environment
Adafruit Libraries written by Limor Fried/Ladyada for Adafruit
*/

#include <Wire.h>
#include "Adafruit_LEDBackpack.h"
#include "Adafruit_GFX.h"
```

```
Adafruit_BicolorMatrix matrix = Adafruit_BicolorMatrix();

void setup() {
  matrix.begin(0x70);
}

void loop() {
EyesCenter();
delay (500);
EyesRight();
delay (500);
EyesLeft();
delay (500);
EyesFlash();
delay(500);
}

/*

  ----------------------------------------------------------
  EyesCenter()
  EyesLeft()
  EyesRight()
  EyesFlash()
  ----------------------------------------------------------
  */

void EyesCenter() {
  matrix.clear();
  matrix.fillCircle(4,3,1, LED_RED);
  matrix.writeDisplay();
}

void EyesLeft() {
  matrix.clear();
  matrix.fillCircle(6,3,1, LED_RED);
  matrix.writeDisplay();
delay (500);
EyesCenter();
}

void EyesRight() {
  matrix.clear();
```

```
    matrix.fillCircle(2,3,1, LED_RED);
    matrix.writeDisplay();
delay (500);
EyesCenter();
}

void EyesFlash() {
  matrix.clear();
  matrix.fillCircle(3,3,3, LED_YELLOW);
  matrix.writeDisplay();
  delay (500);
  matrix.clear();
  matrix.fillCircle(4,3,1, LED_RED);
  matrix.writeDisplay();
  delay (300);
}
```

Operation

Operation is unchanged from the discrete LED method. As the system is triggered, the electronic pupils move toward the source, and then, after lingering for a few hundred milliseconds, the pupils move toward center. Figure 5-22 shows one incarnation of our android head platform that provides a scaffold for hybrid localization, motorized pan with PIR detectors, and unidirectional microphone elements. While adding the ears constructed of silicone pinch bowls provides better results, the unidirectional microphones, backed by sound-absorbing foam, are still a big improvement over omnidirectional microphones.

Figure 5-22a shows mounting of the modified Grove microphone module with silicon sealant inside the vinyl mannequin head. Figure 5-22b shows the same microphone module from the outside. Figure 5-22c provides an inside view of the left-eye mini 8 × 8 matrix. Note the bent acrylic plate immediately in front of the LED, which increases contrast of the LED display and renders the matrix invisible to the casual observer. Figure 5-22d shows a close-up of the left eye with gaze directed forward.

As shown in Figure 5-22e, we inserted a thick foam partition down the center of the head. This foam absorbs pressure waves, creating an acoustic shadow. Finally, Figure 5-22f shows the vinyl support structure head-on. Although the ear microphone modules are just visible, the LED matrices are hidden from casual view by red acrylic plates.

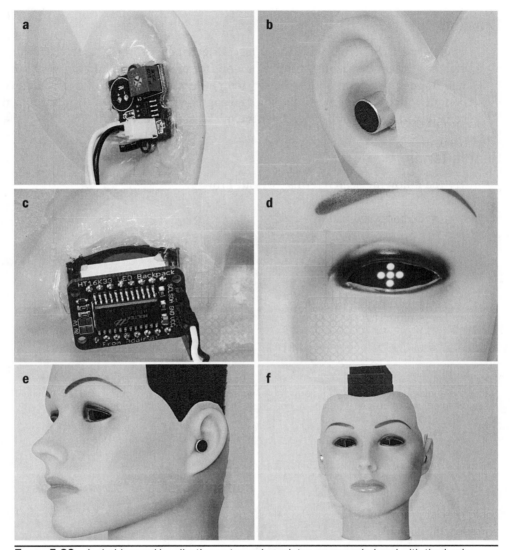

FIGURE 5-22   Android sound-localization setup using a latex mannequin head with the back removed to show project details.

Gremlins

The greatest challenge with the projects described in this chapter is finding or creating the proper acoustic environment, free of reflections and ambient noise. If you happen to live in a glass house, then at least toss a blanket over whatever is behind your experiment. And if you're designing an android for the outdoors, then consider adding a windscreen over the microphone elements.

Search Terms

Try the following search terms for your browser:

- Sound localization
- Sound control
- Soundpoofing
- Electret microphone
- Sensor fusion
- Operational amplifier
- LED matrix
- I2C

CHAPTER 6

A Heartbeat Away

One of the rules for employees of the Disneyland theme parks is that character costumes stay on while the worker is on the set, no matter what. Apparently, the fear is that the sight of a human head protruding from the body of Pluto or Darth Vader would shatter the spell of the magic kingdom, plunging the enterprise into bankruptcy. Whether this is the case or not, there is something abhorrent about the scenes of *StarTrek Next Generation* in which Commander Data's positronic brain is exposed by popping off a hairy head plate. Paradoxically, these scenes serve to remind the viewer that Data is an android and not to lapse into the illusion that he's human. Ideally, diagnosing problems in an android would involve the same methods used on humans. For example, by measuring body temperature, a fever could indicate faulty system operation. Or synthesized heart and breath sounds could be used to indicate drive-system status and efficiency, obviating the need to "pop the top." A potential benefit of blurring the distinction between human physiology and android mechatronics is better human-machine bonding. Think about it, we expect someone who just ran up a flight of stairs to be at least slightly out of breath.

Of course, from a practical perspective, there's value in creating physiologically correct androids that can serve as the equivalent of flight simulators for medics and clinicians in training. Need to train clinicians on how to handle trauma patients? Just rip off a limb from an android and see what happens to the heart rates of both student and android.

Whether to maximize our comfort level with androids, provide a human interface for indicating problems with internal systems, or create a clinical teaching platform, there's value in enabling androids to mimic our physiologic signals. To this end, we'll explore the use of acoustic feedback from heart and breath sounds as signs of the operating status of an android. We'll use a pair of MP3 shields, a pair of Arduino Unos, a few Hall effect sensors, and a pair of 1-W audio amps to create realistic heart and breath sounds that respond to behavior triggers in the internal or external environments.

Biological Basis

The human chest cavity, or thorax, is a packed, busy place. There's the unrelenting contraction and relaxation of the heart and diaphragm and the resulting movement and oxygenation of blood. As we reviewed in Chapter 3, the heat generated by all this metabolic activity must be dissipated to maintain a constant core temperature. In addition to heat, this activity results in a variety of biologically significant sounds and vibrations.

Heart Sounds

When someone sits back in an open field, gazing up lazily at the stars, the motion of her heart produces a familiar "lub-dub" at a regular rate and rhythm. If that same person is thrown into the air by the force of a nearby meteorite plowing into the earth, however, her heart goes into overdrive. Blood volume and pressure skyrocket, slamming the heart valves open and shut. And the diminutive "lub-dub" is replaced with a cacophony that puts downtown Tokyo at rush hour to shame.

A trained clinician can listen to all this activity and diagnose conditions ranging from pneumonia and pneumothorax to mitral regurgitation and aortic stenosis. For example, when one of the four valves of the heart is damaged and won't close completely, the resulting turbulence due to backflow of blood through the valve (regurgitation) results in characteristic vibrations. And these vibrations make their way to the chest wall, where they can be heard with a stethoscope.

Moreover, because of the way the heart is situated in the chest, vibrations from different parts of the heart are heard best at specific locations on the chest wall. For example, the location on the chest wall that corresponds to the vibrations produced by the four heart valves—the mitral, aortic, pulmonic, and tricuspid valves—is shown in Figure 6-1.

Recall that deoxygenated blood flows from the vena cava into the right atrium, through the tricuspid valve into the right ventricle, and through the pulmonic valve to the lungs. Oxygenated blood from the lungs enters the left atrium, passes through the mitral valve into the left ventricle, and then through the aortic valve to the aorta and body. Also note that when facing someone, that person's right is to your left.

In addition, the intensity of specific sounds depends on the position of the person. For example, the abnormal heart sound of mitral regurgitation, which is caused by backflow of blood through a defective mitral valve, is accentuated when a person rolls onto his left side. Because this maneuver doesn't increase the intensity of regurgitation murmurs from the other three valves, this is a way of differentiating the murmur of mitral regurgitation from other regurgitation murmurs.

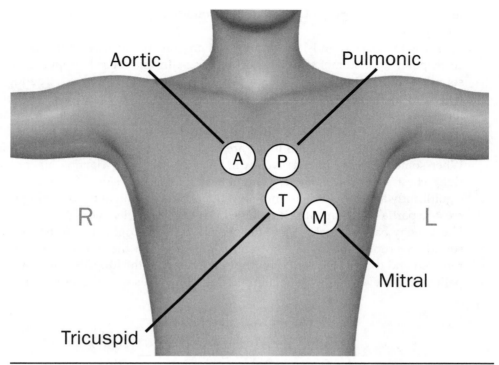

Aortic

Pulmonic

A   P

T

R   M   L

Mitral

Tricuspid

FIGURE 6-1   Locations on the chest where the four heart valves are heard best. Note that the subject's left is to the examiner's right.

The takeaway from this simplified view of biological complexity is that there are thousands of different heart and lung sound presentations. As a result, these sounds have a potential to provide a rich signaling environment for the state of an android—it's simply a matter of mapping human clinical presentations with android system status.

Lung Sounds

Lungs and the sounds they produce are much simpler than the heart and heart sounds. Anatomically, the lungs consist of five separate lobes, three on the right and two on the left. When the diaphragm contracts, the lungs expand to fill the partial vacuum, and air is pulled into the lungs in the process. When the diaphragm relaxes, the elasticity of the muscles in the rib cage and gravity push air out of the lungs. Rapid, deep breathing involves contraction of the muscles of the rib cage to expel air from the lungs.

Even when we're at rest, the movement of air in and out of the lungs creates "whooshing" sounds reminiscent of air rushing through a tunnel. When the rate and/or depth of breathing increase, the "wind" whips around structures in the lung, creating a variety of characteristic sounds depending on the condition of the airway.

As with heart sounds, there are locations on the surface of the body that correspond to anatomic structures of the lungs. In this regard, mapping of lung sounds is much simpler and more obvious, as shown in Figure 6-2. The tops, or apices, of the lungs are heard best near the shoulder blades, and the bases of the lungs are heard best several inches lower. Clinicians normally listen to areas between the base and apex, but this simplified map will do for our purposes.

As "elastic gas bags," the lungs are fairly boring—until something happens. When someone is, say, poked severely in the right chest with a katana, air enters the right chest cavity, creating a pneumothorax. The left side of the chest isn't significantly affected—at least initially. Because contraction of the diaphragm can't create a partial vacuum with air rushing into the chest cavity, the right lung deflates like a soggy balloon. Listening to the chest activity, the right side of the chest will sound cavernous and relatively quiet compared with the left. If the wound is repaired and the air removed from the space between the lung and chest wall, the patient may develop pneumonia in the right lung. This would present initially as

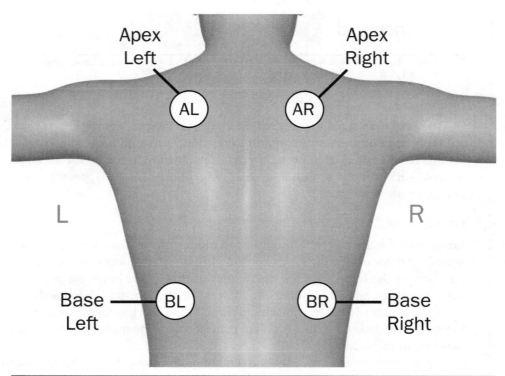

FIGURE 6-2   Locations on the back where sounds from the apex and base of the lungs are heard best.

crackles from fluid buildup near the base of the lung. Of course, this progression of findings assumes the person doesn't die from blood loss.

By the way, an ear pressed against the chest works just as well as a stethoscope, but the cold medical instrument is more hygienic and less invasive. Besides, doctors need something to drape over their shoulders to impress the barista at the local coffee shop.

Relevance to Android Designs

Endowing an android with arbitrary physiologic sounds is akin to the tin man's ticking heart in the *Wizard of Oz*. If your goal is simply to have your android generate monotonous heart sounds, then looping an MP3 file through a small speaker system in the body of the android will do. However, arbitrary sounds aren't going to convey excitement or other emotional cues that humans need to feel connected.

Besides, if you're going through the trouble of decking out your android with a sound system, it might as well have diagnostic capabilities. Need to check on the battery status? Just listen in to the heart rate. A normal, strong heart sound with regular rate and rhythm indicates that all is well. A slower than normal heart rate (bradycardia) can signify that it's time to replace the lithium-ion cells. A faster than normal heart rate (tachycardia) and breathing rate (tachypnea) after light activity could signal that it's time to replace the bearings in the drive system.

Of course, to complete the illusion of a breathing entity, you'll need to install actuators to raise and lower the chest wall in unison with the breath sounds. With the heart, there's the issue of perceptible pulses at various points on the body. Moving the chest wall will have to wait, but we will look into creating lifelike pulses in Chapter 7.

Heart Sound Simulators

There are four main challenges associated with providing an android with an acoustic equivalent of a heart:

- Obtaining or synthesizing clinically correct sounds
- Sensing the location of the stethoscope on the chest wall
- Sensing the orientation and/or activity level of the android
- Playing the appropriate sound

We've taken care of the first challenge for you. You'll find the required sounds online, along with the source code. For the second challenge, sensing the location of the stethoscope, we've used various approaches with good results. One is to attach a small rare-earth magnet to the stethoscope body and trigger either Hall effect sensors or reed relay switches attached to the inner side of the android exoskeleton. Another is to drill pinholes through the exoskeleton and used infrared (IR) light-emitting diodes (LEDs) and photodiodes to detect the presence of a stethoscope on specific areas of the exoskeleton. Capacitive sensing also works, as long as the exoskeleton is nonconductive and no more than a few millimeters thick. The most direct approach is to mount normally open tact switches on the inside of the exoskeleton, with holes in the exoskeleton for the actuators. We'll use Hall effect sensors here.

Determining the orientation of the android body is easily performed with the aid of a simple mercury tilt switch, solid-state tilt sensor, or accelerometer. The latter two have the advantage of operating in multiple planes and that the responses can be modified in software. A mercury tilt sensor, while the least expensive solution in terms of cost and microcontroller input pins, must be physically remounted to alter the directional response. For demonstration purposes, we'll use a mercury tilt sensor.

Playing the appropriate sounds is simplified by the availability of inexpensive MP3 shields for the Arduino. But there's still the challenge of playing the sounds so that they can be heard. The simplest approach, and one that we've used with groups, is to route the audio to an external amplifier and speaker system. However, external audio diminishes the illusion of a real patient. We've used miniature speakers and perforations in the exoskeleton as well as surface transducers with excellent results. For the greatest impact, the audio should be available at the exoskeleton or chest wall and capable of being heard through a stethoscope. We'll use a surface transducer attached directly to the exoskeleton wall.

As you read through the following projects, remember that you have multiple implementation options depending on your budget and the purpose of your android. If you want to learn more about accelerometers, for example, then ditch the retro mercury switch and have at it.

Basic Heart Sound Simulator

In this first project, we'll create the hardware and software infrastructure capable of reproducing physiologically correct heart sounds on the chest wall or exoskeleton of your android. We'll also include the hardware needed for a more advanced heart sound simulator.

Bill of Materials

We've opted to use a sheet of carbon fiber as the build platform, ready for bolting onto a robot. An equally affordable demonstration platform is an $8 plastic mannequin torso clothes hanger, shown front and back with components attached in Figure 6-3. Barring an anatomically correct android, feel free to use anything from a plastic bucket to a cardboard box to hold the sensors and transducers.

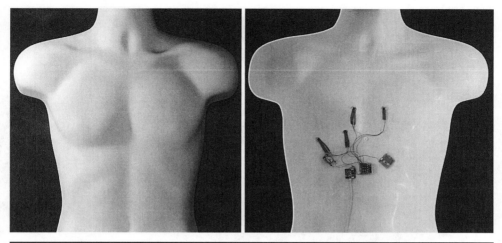

Figure 6-3 Plastic mannequin torso front (*left*) and rear (*right*) showing placement of heart sound simulator components.

The key components for this project, shown in Figure 6-4, include

- Arduino Uno
- 5-V direct-current (dc) power supply
- Sparkfun MP3 player shield
- Isolation circuit (optional; see text)
- 1-GB microSD, FAT 32, with MP3 sound files
- TPA2005D1 1.6-W audio amplifier or equivalent
- 15-kΩ, 1/8- or 1/4-W resistors (2)
- Sparkfun small surface transducer
- Optek Technology OH090U Hall effect sensors (4)
- Mercury tilt sensor or equivalent
- Thin, nonmetallic mounting surface (mannequin optional)
- Small rare-earth magnet
- Stethoscope with aluminum head (optional)

The Sparkfun MP3 player shield is one several capable and affordable MP3 shields for the Arduino on the market. We chose this shield in part because of the onboard microSD drive, the ability to access hundreds of MP3 files through software (as opposed to hardwired pins), and in part because of the excellent library developed by Bill Porter. The shield also has a stereo headphone jack, so you can skip the speakers, surface transducer, and stethoscope if those don't fit your budget.

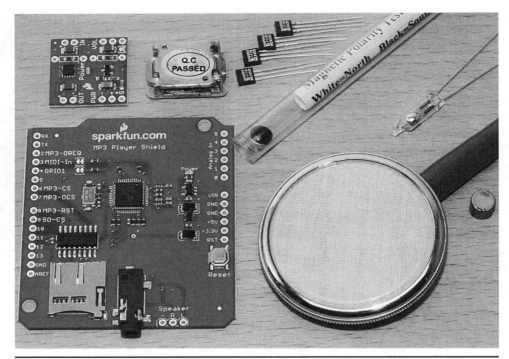

Figure 6-4   Main components of the heart sound simulators.

The TPA2005D1, available on a breakout board from Sparkfun, is a 1.4-W class D audio amplifier. We chose this amp over a generic LM386 amp because the fully floating input can be interfaced directly with the MP3 player shield, and the amp works on a 5-Vdc supply.

The Optek Technology OH090U Hall effect sensor is an inexpensive generic digital sensor available in a standard three-lead Session Initiation Protocol (SIP) case. The nonlatching sensor switches on at 90 G and off at 65 G—meaning that a typical rare-earth magnet has a working distance of about 2 cm. The three-pin device requires a 500- to 860-Ω resistor from the digital output pin to the 5-V supply.

The surface transducer is a speaker coil and magnet designed to use whatever it's attached to as a speaker cone. The surface transducer works on a desktop, sheet of plastic, or any other thin and stiff surface. As an alternative, multiple, small speakers can be mounted on the inside of the exoskeleton of your android. You'll need to make small perforations in the exoskeleton so that the speakers can move air.

Circuit

The circuit for the heart sound simulator is shown in Figure 6-5. The MP3 player shield connections are automatic—simply install the headers and plug the shield

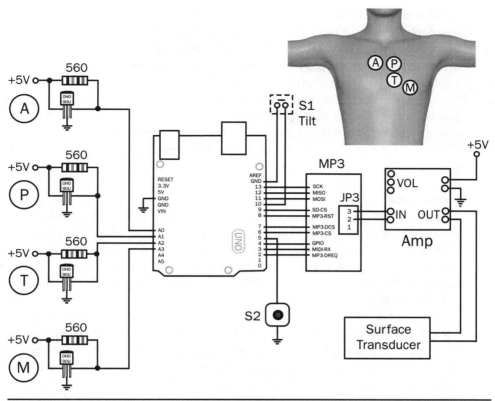

FIGURE 6-5   Schematic of the heart sound simulator.

into the Arduino. Potentially problematic is the connection between the output of the MP3 shield and the input of the audio amplifier. The shield output is fully differential, meaning that both output lines are floating above ground. If you ground one line through an audio amp with one grounded input, then you'll fry the shield. The work-around is to use an amp with a fully differential input, such as the TPA2005D1, or to use an isolation circuit, such as the one suggested by the documentation for the shield. You'll need a pair of 100-Ω resistors, a pair of 10-kΩ resistors, and two 10-μF @ 10-V electrolytic capacitors for the isolation circuit.

If you opt for the TPA2005D1 amp, the interface problems are solved, but you'll need to increase the gain of the amp. The physiologic sounds we're dealing with are in the few hundred hertz range, where the responses of amplifier, MP3 player shield, and surface transducer taper off. To maximize the gain on the TA2005D1, install the two matched 15-kΩ, 1/8- or 1/4-W resistors in the holes provided on the breakout board. The obvious tradeoff for this higher gain is a little more distortion, but it's tolerable.

The outputs of the Hall effect sensors are tied directly to analog pins A0 through A3 of the Arduino. Note the mapping of the analog pins to the chest wall, as shown in the schematic. The reason we're using analog pins is that the serial peripheral interface (SPI)–based MP3 player shield is a pin hog. For non-SPI devices, we have to make due with only digital pins 5 and 10 and the six analog pins.

Not shown in the circuit are the magnets used to activate the Hall effect sensors and the microSD card containing the MP3 files. When you download the files from www.mhprofessional.com/Androids, make certain to download and install all the files onto a microSD card and not simply the ones that correspond to the current project.

Construction

If you're using something that resembles the human torso as an exoskeleton, then follow the illustration in Figure 6-1 for placement of the Hall effect sensors. Otherwise, just keep the relative positions, with the aortic sensor on top, followed, in order, by the pulmonic, tricuspid, and mitral sensors. Leave enough space between sensors to allow the stethoscope to move from one location to the next without overlap.

Figure 6-6 shows the Hall effect sensors together with the amp, surface transducer, and 560-Ω resistor network mounted on a sheet of thin, flexible carbon fiber. For this project, we used wire wrap and hot glue for rapid prototyping and testing. After wrapping the Hall effect sensor leads, we used shrink wrap to keep the leads from shorting and a dab of hot glue to hold everything in place. It doesn't matter whether you mount the Hall effect sensors face up or face down as long as you're consistent.

The four 560-Ω resistors associated with the Hall effect sensors are mounted on a small printed circuit board that serves as a hub for power and ground connections. Not shown in the figure is the MP3 player shield, which connects to the array of sensors through the four wire-wrap posts on the side of the printed circuit board.

You may notice that the leads from the Hall effect sensors are longer than absolutely necessary. The 12-in. lead length is intentional because it supports repurposing of the sensors for follow-up projects without rewiring. However, if you're certain that you won't be using the sensor array elsewhere, then shorten the leads as much as practical to minimize the chances of the leads acting as antennas for potentially interfering signals.

The white disks between the Hall effect sensors and the back of the sheet of carbon fiber are illustrative only so that you can see the relative spacing of the sensors. The disks are the size of a stethoscope head. Note that each Hall effect sensor is located between the center and top of each disk. This is intended to match the location and orientation of the magnet that you'll have to glue to the backside of an aluminum (not steel) stethoscope, as in Figure 6-7. Normally, a stethoscope is held with the audio tube down, an orientation dictated primarily by gravity.

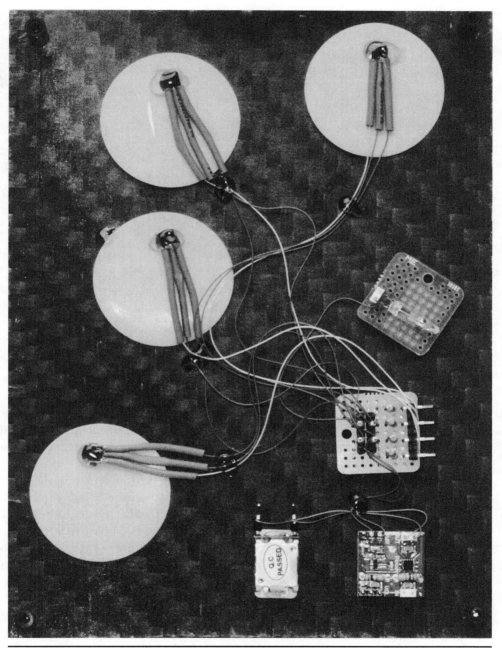

FIGURE 6-6   Heart sound simulator sensors and surface transducer mounted on a sheet of carbon fiber, as seen from the back of the unit. The sheet measures 6 × 8 in.

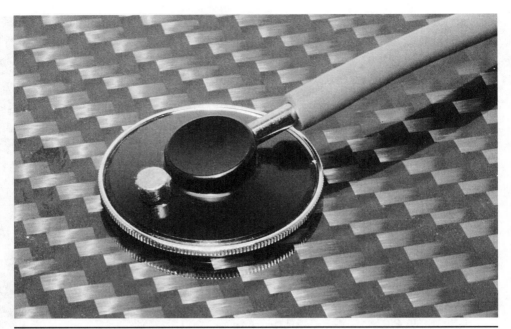

FIGURE 6-7   Rare-earth magnet glued to the back of the aluminum stethoscope head.

Test the polarity of the magnet on a mounted Hall effect sensor before you attach it to the stethoscope head. Once you've verified that the orientation is correct, glue or tape the magnet midway between the center and rim of the stethoscope head, as shown in Figure 6-7. Both tape and hot glue provide instant gratification, as well as easy removal of the magnet later. If you want a semipermanent mount, then use a dab of silicone adhesive.

As shown in Figure 6-6, mount the surface transducer on the backside of the exoskeleton, just below but not too close to the mitral sensor. Before you mount the transducer with hot glue, verify that the amp's 250-kHz pulse-width-modulated (PWM) output doesn't interfere with the nearby sensors. Minimize interference by using a short length of transducer wire, as shown in Figure 6-8. Also note the wire-wrap connections between the surface transducer and amp. You'll have to solder the wire-wrap pins to both the amp and the surface transducer if you want to use the same approach.

Figure 6-8 also shows the two leaded resistors added to the audio amplifier board to increase amplification. If you intend to repurpose the amp on other projects, then you might consider using a dual 50-kΩ potentiometer instead of fixed resistors to provide variable gain. Double-sided tape works well for mounting the small board.

The mercury tilt switch, shown in Figure 6-9, is mounted on a small printed circuit board, which is, in turn, attached, with double-sided tape, to the carbon

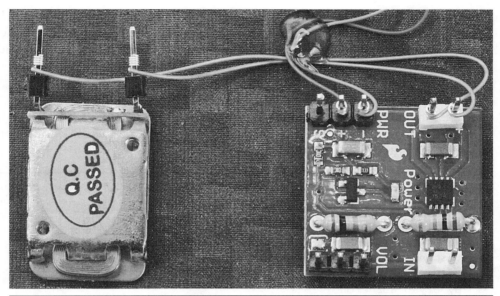

FIGURE 6-8   Details of the amplifier *(right)* and surface transducer *(left)*. Note the 1/4-W leaded gain resistors on the power amplifier board.

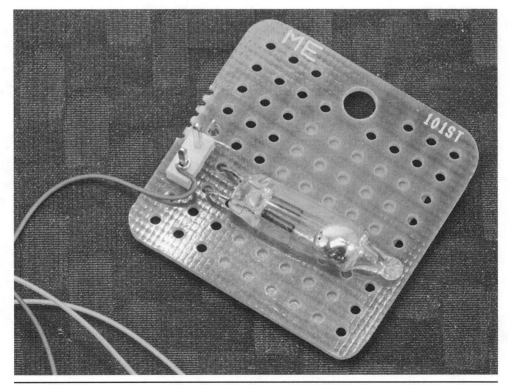

FIGURE 6-9   Close-up of mercury tilt switch. The wire-wrap post is ready for connection to an Arduino.

fiber sheet shown in Figure 6-6. The switch is shown in position, with the bulb slightly down. When the switch and attached carbon fiber sheet are tilted to the left (i.e., counterclockwise in Figure 6-9), the mercury wets the two contacts, grounding the lead to pin D10 of the Arduino.

Code

The Arduino code for the basic heart sound simulator is shown in Listing 6-1. As noted earlier, programming is greatly simplified through the use of Bill Porter's excellent MP3 shield library.

LISTING 6-1   Arduino code for the basic heart sound simulator.

```
/*
Basic Heart Sound Simulator
Androids: Build Your Own Lifelike Robots by Bergeron and Talbot
Listing 6-1. See www.mhprofessional.com/Androids for fully
documented code
By Bryan Bergeron
Arduino 1.0.1 environment
Uses the Sparkfun MP3 Shield Library by Bill Porter
*/

#include <SPI.h>
#include <SdFat.h>
#include <SdFatUtil.h>
#include <SFEMP3Shield.h>

SFEMP3Shield MP3player;

int soundTrack;
int Aortic = 14;
int Pulmonic = 15;
int Tricuspid = 16;
int Mitral = 17;

void setup() {
  pinMode(Aortic,INPUT);
  pinMode(Pulmonic,INPUT);
  pinMode(Tricuspid,INPUT);
  pinMode(Mitral,INPUT);
  MP3player.begin();
  MP3player.SetVolume(15,15);
}
```

```
void loop() {
soundTrack = 1;
if (digitalRead (Pulmonic) == LOW) soundTrack = 2;
if (digitalRead (Tricuspid) == LOW) soundTrack = 3;
if (digitalRead (Mitral) == LOW) soundTrack = 4;
  if (!MP3player.isPlaying()){
      MP3player.playTrack(soundTrack);
      }
  delay(10);
  }
```

Working from top to bottom of the listing, note the multiple libraries loaded, including the standard SPI library. In the constant declaration, the four valve areas are mapped to pins A0 through A3 (pins 14–17). In the setup section, the analog pins are set to digital INPUT pins. Note the SetVolume function, which sets the stereo output volume of the MP3 player shield, with 1 as the maximum volume and 255 as the minimum volume. A setting of 15 gives us some room to increase the volume in a later project. However, if you're having trouble hearing the sounds, then try a setting of 5 or 10.

In the main loop, the soundTrack variable defines the track played by the MP3 player shield through the playTrack() function. Passing a 1 to the function plays track001.mp3, 4 plays track004.mp3, and so on. The heart sound mapping to the supplied audio files is as follows:

- track001.mp3: Normal (aortic)—80 beats/min
- track002.mp3: Normal (pulmonic)—80 beats/min
- track003.mp3: Normal (tricuspid)—80 beats/min
- track004.mp3: Normal (mitral)—80 beats/min

For example, track001.mp3 is the normal heart sound as heard over the aortic area. The rate of contraction and relaxation is 80 beats/min.

The playTrack() function is updated only when the current track has finished playing, as determined by the isPlaying() function. The delay of 10 ms between playback cycles is apparently necessary for proper playback. Fortunately, the delay is not overly obtrusive when the tracks are looped because the recordings start and end with silence.

Operation

To operate the simulator, apply power to the Arduino's input jack, insert the stethoscope earbuds in your ears, and gently place the diaphragm of the stethoscope on the external surface of your platform. You should hear a constant heartbeat. If you're over one of the Hall effect sensors, the sound should change accordingly.

This is a major improvement over a generic loop that plays continuously. Not only are the sounds mapped to the appropriate locations, but by mimicking the pathophysiology of the valves and resulting valve sounds, it's possible to communicate specific information on the status of your android to an external examiner. With this thought in mind, let's add a couple of sensors and a few more lines of code to the project.

Heart Sound Simulator with State Linkages

Given the architecture for normal heart sounds, we'll add abnormal sounds to our library and, more important, link these sounds to sensors that measure internal or external events. These linkages can be software only, assuming that the parameters are available within the Arduino or via serial, I2C, or SPI communications. As noted earlier, this is a software upgrade. You've already installed the hardware infrastructure.

Code

The Arduino code for the heart sound simulator with state linkages is shown in Listing 6-2. As with the basic simulator, all the MP3 files on the McGraw-Hill website should be loaded onto the microSD memory card in the MP3 player shield. The shield should contain sounds of a mitral regurgitation murmur based on the following mapping:

- `track013.mp3`: Mitral regurgitation (aortic)—68 beats/min
- `track014.mp3`: Mitral regurgitation (pulmonic)—68 beats/min
- `track015.mp3`: Mitral regurgitation (tricuspid)—68 beats/min
- `track016.mp3`: Mitral regurgitation (mitral)—68 beats/min

In addition, there are state mappings reflected in the playing of mitral regurgitation sounds and the volume of the sounds.

LISTING 6-2   Arduino code for heart sound simulator with state linkages.

```
/*
Heart Sound Simulator With State Linkages
Androids: Build Your Own Lifelike Robots by Bergeron and Talbot
Listing 6-2. See www.mhprofessional.com/Androids for fully
documented code
Arduino 1.0.1 environment
Uses the Sparkfun MP3 Shield Library by Bill Porter
*/

#include <SPI.h>
```

```
#include <SdFat.h>
#include <SdFatUtil.h>
#include <SFEMP3Shield.h>

SFEMP3Shield MP3player;

int soundTrack;
int Damage_1 = 5;
int LeftTilt = 10;
int Aortic = 14;
int Pulmonic = 15;
int Tricuspid = 16;
int Mitral = 17;

void setup() {
  pinMode(Damage_1,INPUT);
  pinMode(LeftTilt,INPUT);
  pinMode(Aortic,INPUT);
  pinMode(Pulmonic,INPUT);
  pinMode(Tricuspid,INPUT);
  pinMode(Mitral,INPUT);
  digitalWrite(Damage_1,HIGH);
  digitalWrite(LeftTilt,HIGH);
  MP3player.begin();
  MP3player.SetVolume(15,15);
}

void loop() {
soundTrack = 1;
  if (digitalRead (Pulmonic) == LOW) soundTrack = 2;
  if (digitalRead (Tricuspid) == LOW) soundTrack = 3;
  if (digitalRead (Mitral) == LOW) soundTrack = 4;
if (digitalRead (Damage_1) == LOW) {
    if (soundTrack <13) soundTrack = soundTrack + 12;
  }
  MP3player.SetVolume(15,15);
  if (digitalRead (LeftTilt) == LOW) {
if((soundTrack >12) & (soundTrack <17)) {
    MP3player.SetVolume(1,1);
}
  }
  if (!MP3player.isPlaying()){
```

```
        MP3player.playTrack(soundTrack);
      }
   delay(10);
 }
}
```

The code should look familiar. After including the libraries, digital pins 5 and 10 are associated with two new variables, `Damage_1` and `LeftTilt`, respectively. Both pins are defined as inputs with pull-up resistors.

`Damage_1` is a normally open switch that can be used to signify any internal or external event of your choosing, from an elevated temperature to physical contact. When the pin is LOW, `soundTrack` is incremented so that instead the normal heart sounds, the `playTrack()` function plays the location-appropriate sounds of mitral regurgitation.

When `LeftTilt` is LOW, signifying that the mercury tilt switch is tilted to the left, `SetVolume()` is called to set the volume of the MP2 player shield output to maximum, mimicking the physiologic findings of a heart with mitral regurgitation.

The linkages in this simple example are hard coded to the relative position of MP3 file names on the microSD drive. If this code were to be extended to incorporate the thousands of possible clinical presentations, a more compact and maintainable architecture with byte arrays would be appropriate.

Operation

As with the basic heart sound simulator, this simulator continuously plays the sound appropriate for the normal aortic valve area by default. Now, when S2 is closed, the default sound is that of the mitral regurgitation murmur at the location associated with the aortic valve. Moving the stethoscope/magnet over the Hall effect sensors in the pulmonic, tricuspid, and mitral areas results in mitral regurgitation sounds appropriate for those areas. Activating the tilt sensor while the mitral regurgitation sounds are active results in increased amplitude of the murmur sounds.

The logical linkage between the type of damage or sensor input and this particular murmur is up to you. We chose this murmur because the difference between the normal "lub-dub" and the characteristic sound of the murmur should be obvious to even casual listeners.

Lung Sound Simulator

Paralleling the heart sound simulator projects, we'll create a basic lung sound simulator infrastructure and then extend the system with software. Feel free to repurpose the components from the heart sound simulator—easily done if you used the 12-in. leads to the Hall effect sensors. However, note that we'll be combining the heart and lung sound simulators later in this chapter. Also, don't pass up the opportunity to experiment with different sensor technologies, budget and time permitting.

Basic Lung Sound Simulator

Component-wise, the basic lung sound simulator is essentially identical to the basic heart sound simulator. The differences are in the layout of the sensors, following the areas associated with particular lung sounds shown in Figure 6-2, and in the MP3 sound files played.

Bill of Materials

If you've constructed the basic heart sound simulator, then you're set for experimenting with a lung sound simulator. However, as noted earlier, you can take this opportunity to experiment with other sensor technologies.

For example, instead of the digital 0H9009 Hall effect sensors, why not try an analog sensor such as the Allegro A1324? In the absence of a magnetic field, this analog, nonlatching device has an output of 2.5 V. When the north side of a magnet is brought near the face of the sensor, output ramps up to 5 V. When the south side is brought near, the output drops to 0 V.

To use the A1324, you'll need to decide whether to detect an increase or decrease in output and then write a routine that compares the analog signal from the sensors to identify the sensor nearest the magnet. The benefit of this approach is that you needn't hold the stethoscope magnet directly over a Hall effect sensor to activate the nearest sensor. This is particularly applicable to the lung sound simulator because the areas associated with particular lung sounds are more disperse than those associated with specific heart valve sounds.

Circuit

The circuit, shown in Figure 6-10, is virtually identical to that of the basic heart sound simulator. The critical difference, shown in the upper-right corner of the figure, is the placement of the sensors.

Construction

If you're repurposing the components from the preceding project, then simply mount the components on a larger platform, about the size of a human back. A 1-ft^2 sheet of plastic, cardboard, or even carbon fiber also will do.

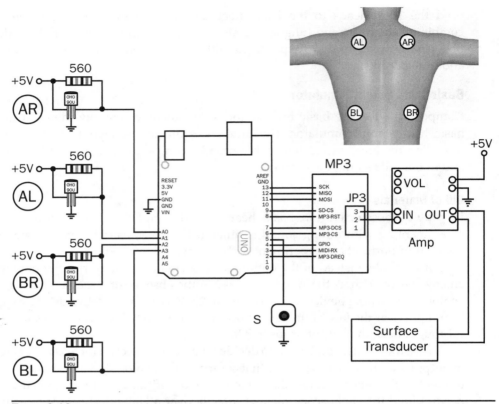

Figure 6-10   Schematic of the basic lung sound simulator.

Code

The Arduino code for the basic lung sound simulator is shown in Listing 6-3. The OH090U Hall effect sensor mappings are as follows:

- A0: Apex left
- A1: Apex right
- A2: Base left
- A3: Base right

The lung sound mapping is

- track017.mp3: Normal apex
- track018.mp3: Normal base

As with the heart sound simulators, the code assumes that all the MP3 files on www.mhprofessional.com/Androids are loaded onto the microSD drive that's plugged into the MP3 player shield.

LISTING 6-3   Arduino code for the basic lung simulator.

```
/*
Basic Lung Sound Simulator
Androids: Build Your Own Lifelike Robots by Bergeron and Talbot
Listing 6-3.
Arduino 1.0.1 environment
Uses the Sparkfun MP3 Shield Library by Bill Porter
*/

#include <SPI.h>
#include <SdFat.h>
#include <SdFatUtil.h>
#include <SFEMP3Shield.h>

SFEMP3Shield MP3player;

int soundTrack = 9;
int UpperLeft = 14;
int UpperRight = 15;
int LowerLeft = 16;
int LowerRight = 17;

void setup() {
pinMode(UpperLeft,INPUT);
pinMode(UpperRight,INPUT);
pinMode(LowerLeft,INPUT);
  pinMode(LowerRight,INPUT);
  MP3player.begin();
  MP3player.SetVolume(15,15);
}

void loop() {
   soundTrack = 17
if (digitalRead (LowerRight) == LOW) soundTrack = 18;
if (digitalRead (LowerLeft) == LOW) soundTrack = 18;

  if (!MP3player.isPlaying()){
```

```
      MP3player.playTrack(soundTrack);
    }
  delay(100);
  }
```

The default value for `soundTrack` corresponds to the sounds associated with the upper lobes of the lungs. Also note that we're only reading the two Hall effect sensors mapped to the base of the lungs. It's useful to have at least four sensors available for the lungs because many clinical presentations aren't bilaterally symmetrical. For example, with a pulmonary infarct, where one lung is collapsed, the default sounds from the left and right apices would not be appropriate.

Operation

Move the stethoscope across the "back" of your platform, whether it resembles a human back or a 5-gallon plastic bucket. The sounds should switch to those associated with the base of the lungs when the stethoscope/magnet is over one of the Hall effect sensors mapped to `LowerLeft` or `LowerRight`.

Lung Sound Simulator with State Linkages

To illustrate how lung sounds can be linked to internal or external events, we'll use a switch to change the state of an input pin on the Arduino. The state change is linked to an MP3 file of coarse crackles in the base of the lungs, a finding typically associated with early pneumonia.

Code

The Arduino code for the lung sound simulator with state linkages is shown in Listing 6-4. The key area in the code is in the main loop, where `Damage_1` is evaluated. If the associated input pin is `LOW` and the stethoscope is over the lower lobes of the lung, then course crackles are played by the MP3 player shield instead of normal base lung sounds.

LISTING 6-4   Arduino code for lung sound simulator with state linkage.

```
/*
Lung Sound Simulator with State Linkages
Androids: Build Your Own Lifelike Robots by Bergeron and Talbot
Listing 6-4. See www.mhprofessional.com/Androids for fully
documented code
Arduino 1.0.1 environment
Uses the Sparkfun MP3 Shield Library by Bill Porter
*/

#include <SPI.h>
```

```
#include <SdFat.h>
#include <SdFatUtil.h>
#include <SFEMP3Shield.h>

SFEMP3Shield MP3player;
int soundTrack = 9;
int Damage_1 = 5;
int UpperLeft = 14;
int UpperRight = 15;
int LowerLeft = 16;
int LowerRight = 17;

void setup() {
  pinMode(Damage_1,INPUT);
  pinMode(UpperLeft,INPUT);
  pinMode(UpperRight,INPUT);
  pinMode(LowerLeft,INPUT);
  pinMode(LowerRight,INPUT);
  digitalWrite(Damage_1,HIGH);
  MP3player.begin();
  MP3player.SetVolume(1,1);
}

void loop() {
soundTrack = 17;
if (digitalRead (LowerRight) == LOW) soundTrack = 18;
if (digitalRead (LowerLeft) == LOW) soundTrack = 18;
if (digitalRead (Damage_1) == LOW) {
    if (soundTrack ==18) soundTrack = 23;
}
  if (!MP3player.isPlaying()){
      MP3player.playTrack(soundTrack);
    }
delay(10);
  }
```

The lung sound mapping is as follows:

- track017.mp3: Normal apex, left and right—16 breaths/min
- track018.mp3: Normal base, left and right—16 breaths/min
- track023.mp3: Coarse crackles base—16 breaths/min

Operation

As with the previous lung sound simulator, operation involves donning a stethoscope/magnet and listening to the "back" of your platform. When `Damage_1` is brought `LOW`, listening at the base of the lungs with the stethoscope/magnet should reveal coarse crackles.

Coordinated Heart and Lung Sound Simulators

The stand-alone simulators presented earlier are meant to ignite your interest in creating physiologically correct signaling. And we've only touched the surface with a single murmur and one abnormal breath sound. We expect you to expand on these simple demonstration programs.

Perhaps the easiest way to expand what we've done thus far is to integrate the heart and lung sound simulators. Integration—even as loosely as defined here—pushes mimicry to a higher level, opening up more possibilities of physiologically correct presentations. Whereas full integration is beyond the scope of this chapter, you can get a taste for the level of complexity involved in the process by constructing the coordinated simulators.

Our goal is to illustrate how even simple integration can enhance the utility of a simulator for both communicating internal state and creating more human-like interfaces. Best of all, if you've followed along thus far, there's little to do other than move the simulators to the same platform and upload the programs listed next.

Bill of Materials

To construct the coordinated simulators, you'll ideally have the previously developed heart and lung sound simulators mounted on a common platform. This will give you a sense of listening to both heart and lung sounds simultaneously. Other than that, you can get by with a pair of normally open switches or, preferably, another Arduino Uno and a few feet of wire.

Circuit

The high-level circuit for the coordinated simulator project, shown in Figure 6-11, consists of the two Arduino-based simulators previously described and a third Arduino that serves as coordinator. Pins A4 and A5 of both simulators are tied to the digital output of the third Arduino or a pair of switches. If you opt for tighter coordination, then you should use a more robust connection, such as SPI.

Code

Following in Listings 6-5 and 6-6 are the modified listings for the heart and lung sound simulators, respectively. Of course, the code is meaningless without information on the sound files on disk. The following files are loaded onto the microSD cards of respective simulators:

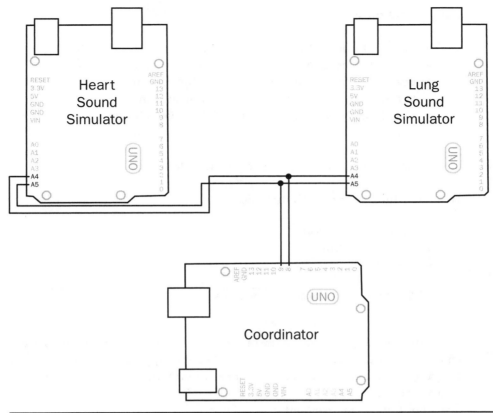

Figure 6-11 Schematic of the coordinated simulator system. Details of the sensors associated with each simulator are hidden for clarity.

Heart sound files:

- `track001.mp3`: Normal (A)—80 beats/min
- `track002.mp3`: Normal (P)—80 beats/min
- `track003.mp3`: Normal (T)—80 beats/min
- `track004.mp3`: Normal (M)—80 beats/min
- `track005.mp3`: Bradycardia (A)—48 beats/min
- `track006.mp3`: Bradycardia (P)—48 beats/min
- `track007.mp3`: Bradycardia (T)—48 beats/min
- `track008.mp3`: Bradycardia (M)—48 beats/min
- `track009.mp3`: Tachycardia (A)—108 beats/min
- `track010.mp3`: Tachycardia (P)—108 beats/min
- `track011.mp3`: Tachycardia (T)—108 beats/min
- `track012.mp3`: Tachycardia (M)—108 beats/min

Lung sound files:

- `track017.mp3`: Normal apex—16 breaths/min
- `track018.mp3`: Normal base—16 breaths/min
- `track019.mp3`: Bradypnea apex—9 breaths/min
- `track020.mp3`: Bradypnea base—9 breaths/min
- `track021.mp3`: Tachypnea apex—24 breaths/min
- `track022.mp3`: Tachypnea base—24 breaths/min<

The activity logic for pins A4 and A5 are as follows:

Heart and lung sounds activity logic:

- 01: Asleep, sedated
- 00: Awake, sitting
- 10: Awake, exercising

LISTING 6-5   Arduino code for coordinated heart sound simulator.

```
/*
Coordinated Heart Sound Simulator With External Activity Link
Androids: Build Your Own Lifelike Robots by Bergeron and Talbot
Listing 6-5. See www.mhprofessional.com/Androids for fully
documented code
Arduino 1.0.1 environment
Uses the Sparkfun MP3 Shield Library by Bill Porter
*/

#include <SPI.h>
#include <SdFat.h>
#include <SdFatUtil.h>
#include <SFEMP3Shield.h>

SFEMP3Shield MP3player;
int soundTrack;
int Aortic = 14;
int Pulmonic = 15;
int Tricuspid = 16;
int Mitral = 17;
int Activity0 = 18;
int Activity00 = 19;
```

```
void setup() {
  pinMode(Activity0, INPUT);
  pinMode(Activity00, INPUT);
  pinMode(Aortic,INPUT);
  pinMode(Pulmonic,INPUT);
  pinMode(Tricuspid,INPUT);
  pinMode(Mitral,INPUT);
  digitalWrite(Activity0,HIGH);
  digitalWrite(Activity00,HIGH);
  MP3player.begin();
  MP3player.SetVolume(15,15);
}

void loop() {
  soundTrack = 1;
  MP3player.SetVolume(15,15);
  if (digitalRead (Pulmonic) == LOW) soundTrack = 2;
  if (digitalRead (Tricuspid) == LOW) soundTrack = 3;
  if (digitalRead (Mitral) == LOW) soundTrack = 4;

if (digitalRead (Activity00) == LOW){
    soundTrack = soundTrack + 8;
}
else if (digitalRead (Activity0) == LOW){
    soundTrack = soundTrack + 4;
}

if (!MP3player.isPlaying()){
MP3player.playTrack(soundTrack);
}
delay(10);
}
```

The key additions to the basic heart sound simulator code are pins A4 and A5, referred to as Activity0 and Activity00, respectively. Both A4 and A5 are defined as INPUT, and their internal pull-up resistors are activated. The variable soundTrack is incremented if either Activity0 or Activity00 are LOW so that the sounds played are slower than normal (Activity0 = +4 increment) or faster than normal (Activity00 = +8 increment). The default is normal rate.

LISTING 6-6   Arduino code for coordinated lung sound simulator.

```
/*
Coordinated Lung Sound Simulator with External Activity Link
Androids: Build Your Own Lifelike Robots by Bergeron and Talbot
Listing 6-6. See www.mhprofessional.com/Androids for fully
documented code
Arduino 1.0.1 environment
Uses the Sparkfun MP3 Shield Library by Bill Porter
*/

#include <SPI.h>
#include <SdFat.h>
#include <SdFatUtil.h>
#include <SFEMP3Shield.h>

SFEMP3Shield MP3player;
int soundTrack;
int UpperLeft = 14;
int UpperRight = 15;
int LowerLeft = 16;
int LowerRight = 17;
int Activity0 = 18;
int Activity00 = 19;

void setup() {
pinMode(Activity0, INPUT);
  pinMode(Activity00, INPUT);
  pinMode(UpperLeft,INPUT);
  pinMode(UpperRight,INPUT);
  pinMode(LowerLeft,INPUT);
  pinMode(LowerRight,INPUT);
  digitalWrite(Activity0,HIGH);
  digitalWrite(Activity00,HIGH);
  MP3player.begin();
  MP3player.SetVolume(1,1);
}

void loop() {
  soundTrack = 17;
  if (digitalRead (LowerRight) == LOW) soundTrack = 18;
  if (digitalRead (LowerLeft) == LOW) soundTrack = 18;
```

```
if (digitalRead (Activity00)){
   soundTrack = soundTrack + 4;
  }
else if (digitalRead (Activity0)){
   soundTrack = soundTrack + 2;
}

  if (!MP3player.isPlaying()){
      MP3player.playTrack(soundTrack);
     }

  delay(10);
 }
```

Mirroring the listing of the coordinated heart sound simulator, pins A4 and A5 are labeled `Activity0` and `Activity00`, respectively. The pins are defined as input ports, and the internal pull-up resistors are activated. The variable `soundTrack` is incremented if either `Activity0` or `Activity00` is LOW so that the lung sounds played are slower than normal (`Activity0 = +2 increment`) or faster than normal (`Activity00 = +4 increment`). The default is normal rate appropriate for the location on the back.

Operation

Pulling pin 18 (A4) LOW on the simulators causes both the heart and lung simulators to play sounds associated with an asleep, sedated person. Conversely, pulling pin 19 (A5) LOW causes the simulators to play sounds associated with exertion and exercise.

From here, it's simply a matter of adding logic for specific presentations, linking the logic to internal or external events, and adding the appropriate sounds to the microSD disk. We've developed synthesizers for this purpose, but you can find example sounds on the Web. Ideally, microcontrollers will eventually be powerful enough to synthesize the sounds directly, allowing full integration of sounds. For now, playing the appropriate MP3 files goes a long way past the status quo.

Gremlins

The greatest challenge with these simulators, and this chapter in general, is staying focused. There are so many options and rabbit holes—you could spend a month evaluating and experimenting with MP3 player shields. And then there's the world of Hall effect sensors. Some are linear, others are digital, some lock, some don't, some require reversal of magnetic field for a reset, and others simply require removal of the magnetic field. The surface transducer is another potential diversion. It's hard to resist testing the device on china plates, the forehead, teeth, Mylar film, and other potential sound radiators.

In short, we went through several iterations of technology before deciding on simply keeping it simple. And perhaps you should do the same—at least at first.

Search Terms

Try the following search terms for your browser:

- Cardiac auscultation
- Pulmonary auscultation
- Cardiac maneuvers
- Heart murmurs
- Hall effect sensor
- Tachycardia
- Bradycardia
- Tachypnea
- Bradypnea

CHAPTER 7

If It Bleeds,
Can We Kill It?

T he most convincing androids depicted in film are not dry, gunmetal-gray mechatronic contraptions with vinyl-covered exoskeletons. They're warm, wet creatures perfused with "blood" that usually looks like something between Elmer's glue and white gravy when their skin is punctured or something is ripped from their torso.

Aside from giving an android a solid, fleshy sound and feel, what's the value of fluid-filled spaces instead of air? As we discussed in Chapter 3, there's value in using fluid transfer of heat from the core to the periphery, where it can be dissipated. There's also the human-interface element of realism. Distended jugular veins emphasize anger, for example. And intense muscular exertion is normally accompanied by distended veins in the forehead, neck, and arms. Also, as with physiologically correct sounds, the nature of a pulse, if present in an android, could be used to communicate, in a natural way, the status of internal systems.

In this chapter you'll learn how to give your android a palpable pulse that can mimic the appropriate response to exercise. We'll also investigate Korotkoff sounds and show you how to integrate a pulse simulator with the heart and lung sound simulators of Chapter 6. The projects require little more than a water pump, motor shield, Arduino, and some flexible polyvinyl chloride (PVC) tubing. You'll need an inexpensive blood pressure cuff and stethoscope to fully appreciate the Korotkoff sounds.

Biological Basis

The human circulatory system is primarily a closed transport system for nutrients and waste. Like the stream in a circular sushi boat buffet system, there's a constant flow of fluid, propelled by the heart, arteries, and skeletal muscle activity, that moves oxygen, proteins, carbohydrates, and fats to where they're needed. This fluid—blood—also moves carbon dioxide and waste products to organs where they can be eliminated. For our purposes, we'll focus on the physical architecture of the circulatory system—the heart and associated blood vessels—and touch on the control mechanisms that determine the flow of blood through the system.

Circulatory System Architecture

As we reviewed in Chapter 6, the heart pumps blood to and from the lungs and back into the systemic circulation that includes the skeletal muscles and internal organs. Figure 7-1 shows the plumbing of the systemic circulatory system. The high-pressure output side of the heart is connected to a large-diameter, thick-walled artery, the aorta. This vessel branches into smaller-diameter, thinner-walled arteries. These, in turn, branch into arterioles and finally to the porous capillaries, where nutrient and waste exchange occurs.

The branching architecture is reversed on the low-pressure return side of the heart, where multiple capillaries coalesce into venules, and multiple venules combine to form small veins. These, in turn, combine in two large-diameter veins, the superior and inferior vena cavae, that return blood to the heart.

Compared with arteries, veins have relatively thin walls with little muscle. Veins also tend to be nearer the surface than arteries, which makes sense from a

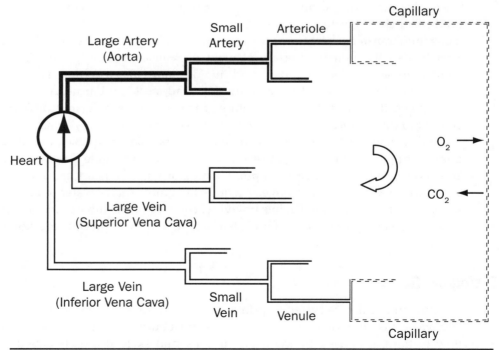

FIGURE 7-1 Simplified view of systemic circulation, with thick-walled, high-pressure arterial source (*above*) and thin-walled, low-pressure venous return (*below*).

survival perspective. Cut a surface vein and you can probably get by with a Band-Aid. Even if you intentionally slit the veins in your wrist, you'd probably have to sit in a tub of warm water for hours to keep clots from plugging the site of injury.

If you really want to end it all, you'll want to cut deep, to the bone, so that you slice open—but don't sever—the radial artery. If you sever the artery cleanly, the muscular wall will contract, sealing the end. Hence the value, in a life-or-death fight, of twisting the knife after you've plunged it into your opponent. It also explains why, after slicing through the aorta while committing seppuku, a Japanese warrior relied on a second to lop off his head. Otherwise, the warrior could writhe in pain for an hour or more, making for a messy, lengthy ceremony.

Blood Flow

Blood flows from the arteries to the veins primarily because of the cyclic contraction and relaxation of the heart. The fourth chamber of our four-chambered heart, the left ventricle, does the heavy lifting. When it contracts (systole), a high-pressure pulse of blood squirts out into the large aorta. When the left ventricle relaxes to refill with blood (diastole), pressure in the aorta dips. Pressure in the cardiovascular system is measured in millimeters of mercury (mmHg), or the height of a column of mercury that would be suspended by the pressure.

One of the reasons that we don't stroke out and die from an endless torrent of high-pressure hammer pulses from the heart by the age of 16 is the elasticity of the arterial system. The thick-walled arteries are critical not only in providing containment of high-pressure fluid, but they also stretch with each pulse of blood from the heart and then rebound when the left ventricle relaxes. This moderates the pressure and keeps it elevated enough to move the blood through the system. Think of the heart as a pulse-width-modulated (PWM) generator and the aorta, arteries, and arterioles as filter capacitors, both decreasing the maximum amplitude of the pulse and providing a lower level but steady direct-current (dc) voltage. Figure 7-2 presents a simplified analogous model of the cardiovascular system.

Starting at the main generator, G1, a PWM source and diode D1 represent the heart. The diode represents the four one-way valves that normally keep blood moving in one direction. Series resistor R1 and capacitor C1 represent the resistance and elasticity of the large aorta. Potentiometers P1 and P2 represent the variable distributed resistance of the smaller arteries and arterioles, and C2 represents the elasticity of these vessels. Resistor R4 represents the significant resistance of the capillaries.

Now, looking at the return or venous side of the circuit, diode D2 represents the numerous one-way valves in the venous system. Potentiometers P3 and P4 represent the variable distributed resistance of the veins and venules. The second generator, G2, which incorporates diode D3, represents the venous pump, the action of the skeletal muscles, and the abdominal pressure created by the diaphragm with each breath.

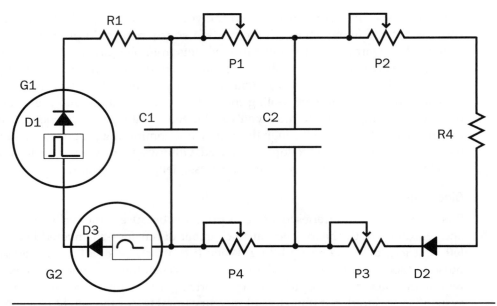

FIGURE 7-2   Electronic equivalent of the cardiovascular system.

Figure 7-3 shows a stylized blood pressure pulse wave in the arteries that parallels what one would expect to see after passing a square pulse through a simple resistor-capacitor (RC) filter. Using electronic equivalents, the systolic pressure is equivalent to peak voltage, and diastolic pressure is equivalent to the dc offset voltage. The heart rate, in beats per minute, is equivalent to the frequency of the pulse train, and the relative time of systole to diastole is equivalent to the duty cycle.

Values shown in the figure are a heart rate of 60 beats/min ("1.0 Hz"), a systolic pressure ("Peak voltage") of 160 mmHg, a diastolic pressure ("Dc offset") of 70 mmHg, and a systolic-to-diastolic time ratio of roughly 1:2 ("Duty cycle = 30 percent"). An actual pulse wave is a little more complicated than the one depicted in the figure largely because of internal reflections of the pulse pressure wave.

The heart of an average adult human sitting in a cubical in an office building pumps about 5 liters of blood per minute. This is roughly the total volume of blood in the body—think 10 pints of Ben and Jerry's Chunky Monkey. The heart of that same human running all-out, or watching the scene from *Alien* when the creature pops through the chest wall, is about 35 liters of blood per minute. Think 70 pints of Chunky Monkey. The heart accomplishes this feat by increasing the frequency of contraction and by propelling more blood with each contraction, that is, by increasing both the frequency and duty cycle of the cardiac PWM generator.

Top marathon runners and superheroes have resting pulse rates in the range of 40 and 50 beats/min. The remaining 99 percent of us have resting pulses in the

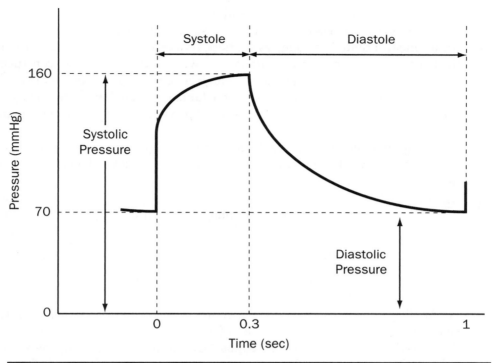

FIGURE 7-3 Stylized pulse pressure wave in the arteries.

range of 60 to 90 beats/min. During extreme exertion, the hearts of those marathon runners might tack up to 180 beats/min. Cubicle rats might hit 130 or 140 beats/min for a few minutes before exhaustion. A rough estimate of maximum theoretical heart rate is provided by the formula

$$\text{Heart rate maximum} = 220 - \text{age in years}$$

So, if your android is supposed to be 24 years old, it's maximum heart rate is about $220 - 24 = 196$ beats/min. But short of trying to escape from a wild lion or bench pressing that extra 100 lb to impress the girls in the gym, few humans push it hard enough to reach their maximum heart rate.

Exercise alters the total amount of blood pumped by the heart, as well as the distribution of that blood flow. At rest, about 20 percent of total blood flow is through the biceps, quadriceps, and other skeletal muscles. During hand-to-hand combat, that total can increase to 85 percent or more. One-hundred percent skeletal muscle flow isn't possible simply because the brain, heart, and diaphragm, as components of the primary life-support system, need blood to keep functioning.

The redistribution of blood flow is ultimately determined by the muscle tone of the arterial system. Relaxation of the muscles in the arterial walls results in widening of the lumen or central passageway of arteries (vasodilation). Conversely, increasing the tension of these muscles results in narrowing of the lumen (vasoconstriction). Because the resistance to blood flow through an artery is proportional to $1/(\text{lumen radius})^4$, small changes in the radius of the lumen result in large flow changes. For example, decreasing the lumen radius by 50 percent increases resistance to blood flow by a factor of 16.

Using an analogy with Ohm's law, voltage is equivalent to blood pressure, current is equivalent to blood flow, and resistance to electron flow is equivalent to the resistance of a vessel to blood flow. That is,

$$\text{Current} = \text{voltage}/\text{resistance}$$

and

$$\text{Blood flow} = \text{pressure}/\text{resistance}$$

Another notable parallel with electronics is that just as copper wire is rated at so many ohms per foot, the resistance of an artery is proportional to the length of the artery. Double the distance blood must flow through an artery, and the resistance to blood flow is doubled.

Up to this point, we've focused on the arterial side of the circulatory system largely because that's where the action is. Veins are thin-walled, passive vessels with one-way valves at various intervals to prevent backflow of blood. Most of the blood at any moment is in the veins, en route to the heart. This lazy river of fluid is helped downstream by the action of the venous pump described earlier.

Peripheral Pulses

The pulse waves generated by the heart, even if moderated by the elasticity of the arterial system, keep blood moving throughout the body. A quick method of checking whether a person's cardiovascular system is working to spec is to check pulses in the periphery, distant from the heart. Although there are various electronic instruments up to the task, checking the intensity and rate of the pulse is as simple as compressing an artery against a bone with the tips of the index and middle fingers.

In the movies, the absence of a pulse in a victim's neck (carotid artery pulse) is a sure sign of death. Although this is usually true, in the real world, the strengths of the pulses at the wrist (radial artery pulse) and ankle joint (posterior tibial artery) are also critical values. A heart that's taken a bullet may be strong enough to push blood up a few inches to the large carotid artery but not strong enough to push blood down several feet of small-radius, relatively high-resistance artery to the ankle.

Korotkoff Sounds

You may be wondering how pressure is determined in the Ohm's law equivalent of the blood flow equation discussed earlier. The most accurate method involves inserting a pressure probe directly into the arterial system. Because this method is painful, expensive, and involves large-bore needles, most people opt for a less accurate approach that involves the sounds made by turbulent blood flow.

If you worked with the mitral regurgitation murmur in Chapter 6, you know the "whooshing" sound made by turbulent blood passing through an incompetent valve. You can make that same sound by compressing an artery with your fingertips. Compress the artery completely so that blood is cut off, and the turbulence sound disappears. If you listen "downstream" from the artery with a stethoscope, you should hear the changes in sound as you vary the pressure on the artery.

Whereas fingers work fine for producing Korotkoff sounds (named after Dr. Nikolai Korotkoff), unless your fingers are calibrated in millimeters of mercury, you'll have to use a blood pressure cuff and pressure meter for accurate pressure measurements. Start with the cuff fully inflated over an artery, which completely compresses the artery. Slowly release the air in the cuff, and note the pressure at which you first hear a faint snapping sound. This is systolic blood pressure. Continue releasing pressure until the "whooshing" sound disappears. This is the diastolic blood pressure. Blood pressure is typically reported as systolic over diastolic, such as 135/80 mmHg.

Relevance to Android Designs

If you want to realistically simulate human physiologic responses, then you've got to incorporate broad-brushstroke factors such as the pulsatile nature of blood flow. As we discussed earlier, there's value in pure mimicry, as well as in communicating internal functioning to an outside observer. For example, the strength of ankle pulses can be used to signal the integrity of an underlying mechatronic system—say, the hydraulics.

The approach we've taken in the projects described herein is to use a hydraulic system that leverages much of the work we did in Chapter 3. If you didn't work through the projects described in that chapter, now would be a good time to review the discussions on materials and construction. If you have an aversion to working with water or simply don't have the resources to invest in tubing and a pump, then there are all-electronic alternatives.

For example, if your goal is to simply provide a tactile pulse at specific points on your android, then a dc solenoid is an affordable alternative. We've had good luck with the line of solenoids from Robot Geek, which have threaded mounting holes as well as threaded holes on each end of the plunger. With minor modification to the code that we provide later, your robot can have all-electronic pulses. Simply mount the solenoids so that pressure is exerted outward with each beat. In tight spaces, you'll need to build the equivalent of a lever to transform side-to-side motion of the plunger to up-down motion at the exoskeleton of your android. The downside is the "clickety-clack" noise from the solenoid, which tends to be amplified when it's attached to a plastic or lightweight metal body.

Similarly, you could provide Korotkoff sounds by repurposing the MP3 player shield used in Chapter 6. You'd need to tie in the sound-file selection with pressure sensors over the areas that you intend to "compress." However, at this point, the complexity of an all-electronic version of a cardiac pulse simulator becomes a limiting factor. So if you want the full simulator experience, go with the flow.

Cardiac Pulse Simulator

We're going to build a cardiac pulse simulator with peripheral pulses and the ability to elicit Korotkoff sounds with a standard blood pressure cuff and stethoscope. We'll then move to modeling the exercise response, followed by integration of the pulse generator with the heart and lung sound simulators we developed in Chapter 6. Other than setting up an Arduino with a motor shield, most of the work involves plumbing and keeping the electronics dry.

Bill of Materials

To construct the cardiac pulse simulator, you'll need the following:

- 12-Vdc submersible water pump
- 5 ft of ½-in. inside diameter (ID) clear PVC tubing
- 12 in. of ¾-in. ID clear shrink-wrap tubing
- 11-qt plastic dishpan
- Arduino Uno or equivalent
- Duct tape
- Small plastic clip
- 12-V battery
- Motor control shield or N-channel metal-oxide semiconductor field-effect transistor (MOSFET) rated at 2 A
- 3½- × 12-in. PVC pipe or equivalent
- Stethoscope (optional)
- Blood pressure cuff (optional)

The heart of this project is the submersible water pump, shown in Figure 7-4. As in Chapter 3, we use a Rule submersible 12-Vdc @ 1.2 A bilge pump rated at 350 gal/h. This is the minimum-capacity pump you should consider. Lift capacity isn't listed in the product documentation, but we estimate that it's at least 6 ft. Also shown in Figure 7-4 is Sparkfun motor shield, based on N-channel MOSFETS, to provide high-current output to drive the pump.

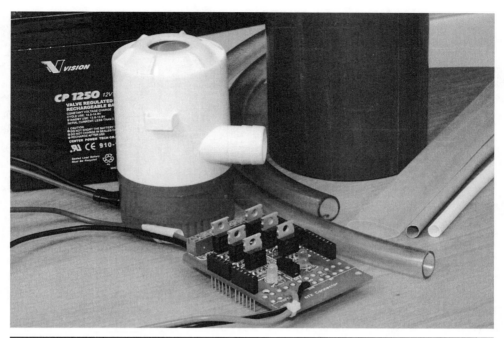

FIGURE 7-4   Major components used to construct and operate the cardiac pulse generator.

Circuit

The circuit, shown in Figure 7-5, consists of an Arduino, 12-Vdc pump, *N*-channel MOSFET interface, and 12-V battery. A benefit of using the Sparkfun motor shield instead of a single *N*-channel MOSFET is that the 12-V battery supplies both the pump and the Arduino. You don't want to be tied down to an outlet and brick power supply when working around water. If you decide to use a single MOSFET device, then use one that has built-in diode protection. Otherwise, you'll have to supply external diodes to protect the MOSFET from spikes generated by the pump.

Construction

Construction consists of creating an 8-in.-long thin-walled section of tubing between two lengths of the thick PVC tubing and connecting one end to the pump and the other to the water container. Cut the 5-ft length of PVC tubing in half, and reconnect the end by inserting each cut end about 2 in. into the 12-in. heat-shrink tubing. Use a heat gun to shrink the tubing onto the PVC tubing. After it cools, mount the tubing section onto the PVC pipe using tape or tie wraps on the thick ½-in. ID PVC tubing only, as shown in Figure 7-6.

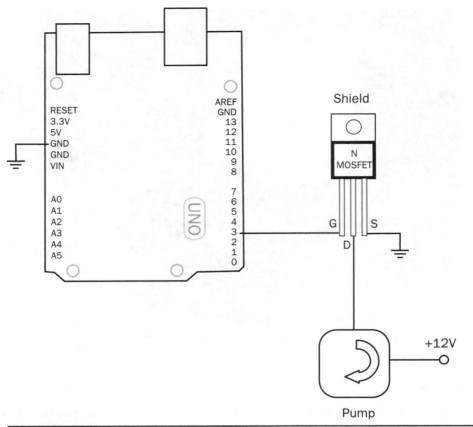

FIGURE 7-5   Cardiac pulse simulator circuit.

FIGURE 7-6   Thin-walled shrink-wrap connector between lengths of ½-in. ID PVC tubing mounted on 3.5-in. PVC pipe.

Connect one free end of the PVC tubing to the pump using an adapter or duct tape. In our case, the tubing was a tight fit with the pump connector, and a few inches of duct tape were enough to keep everything in place. Use a clip to secure the return end of the tubing so that it doesn't jet around the pan and cause a mess once the pump is energized.

Code

The Arduino code for the cardiac pulse simulator is shown in Listing 7-1.

LISTING 7-1   Arduino code for the cardiac pulse simulator.

```
/*
Cardiac Pulse Simulator
Androids: Build Your Own Lifelike Robots by Bergeron and Talbot
Listing 7-1. See www.mhprofessional.com/Androids for fully
documented code
Arduino 1.0.1 environment
*/

constintpumpPin = 3;
constintsystolicSpeed = 255;
constintdiastolicSpeed = 100;
constintpulseRate = 60;
constintdutyCycle = 30;

doublesystolicDelay;
doublediastolicDelay;
doubletotalMS;

void setup() {
totalMS = (60000/pulseRate);
systolicDelay = (dutyCycle * totalMS)/100;
diastolicDelay = totalMS - systolicDelay;
}

void loop() {
analogWrite(pumpPin,systolicSpeed);
delay (systolicDelay);
analogWrite(pumpPin,diastolicSpeed);
delay (diastolicDelay);
}
```

Working from the top, the characteristics of the cardiac pulse wave are defined. `SystolicSpeed` is the maximum motor speed during high-pressure systole, and `DiastolicSpeed` is the motor speed during low-pressure disatole. Because there isn't a second pump to simulate the elastic rebound of the aorta and other arteries, the pump is kept running during diastole. The values given for `systolicSpeed` and `diastolicSpeed` result in a systolic pressure of 160 and 40 mmHg, respectively, assuming that you use the same pump and tubing described earlier.

The constants `pulseRate` and `dutyCycle` define the pulse in beats per minute and the duty cycle in percent, respectively. During setup, `systolicDelay` and `diastolicDelay` are assigned values in milliseconds for the delay function. The main loop continuously writes the systolic speed and then diastolic speed to the pump. Pump speed doesn't change during the `delay()` calls.

Notice that the variable for total milliseconds per cycle, `totalMS`, and the two delay variables, `systolicDelay` and `diastolicDelay`, are defined as doubles. If you keep pulse rate at 60 beats/min or above, you can use integer variables. Otherwise, the program will crash with an overflow error during setup.

Operation

Start by clearing a spot on your kitchen countertop near the sink to debug the system. Once you're satisfied that the connections are solid, you can return to your workbench, but keep a few towels handy. Fill the dishpan to about 4 in., and submerge the pump. Next, verify that the return end of the PVC tubing is still secured by the plastic clip. The section of shrink-wrap tubing that's attached to the 3.5-in. PVC pipe should be immediately adjacent to the pan at the same level as the pump. Last, connect the 12-V battery.

You should see the shrink-wrap section of tubing pulsate once a second. You should also hear the pitch of the pump sound change from high to low as the speed decreases from systolic to diastolic settings. Next, place your index and middle fingers across the shrink-wrap tubing, and press down gently, as shown in Figure 7-7. You should feel a pulse, much like the pulse at your carotid.

Next, put on your stethoscope. Assuming that water is flowing into the tubing from left to right, put your left index finger across the middle of the tubing, and gently press down while listening, just to the right of your finger, with the diaphragm of the stethoscope (Figure 7-8). You should hear the "whooshing" of Korotkoff sounds as you move from full patency to a closed lumen and back again. We found the sounds much more prominent and easily distinguished compared with sounds typically heard during a routine blood pressure test.

If you have access to a blood pressure cuff, then, with the pump still running, attach the cuff over the left end of shink-wrapped tubing (again, assuming flow is from left to right). Line up the "artery" arrow on the cuff with the tubing, and secure the Velcro. Quickly inflate the cuff so that flow ceases. Then, while listening with the diaphragm of the stethoscope over the exposed right edge of shrink-wrap

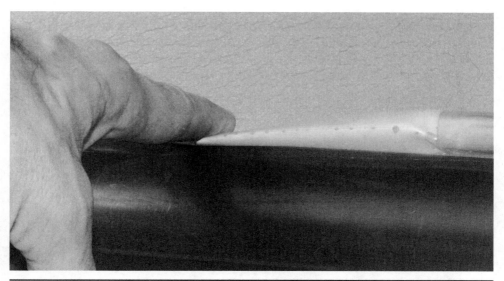

FIGURE 7-7   Feeling for a pulse by compressing the thin shrink-wrap tubing toward the 3.5-in. pipe base.

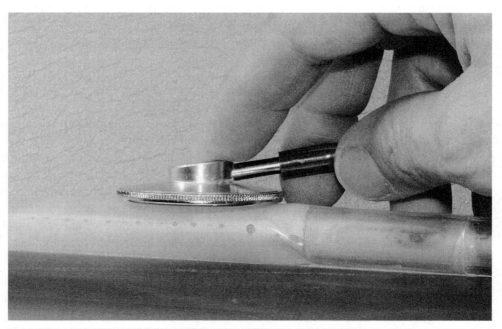

FIGURE 7-8   Listening for Korotkoff sounds downstream from the compressed shrink-wrap tubing.

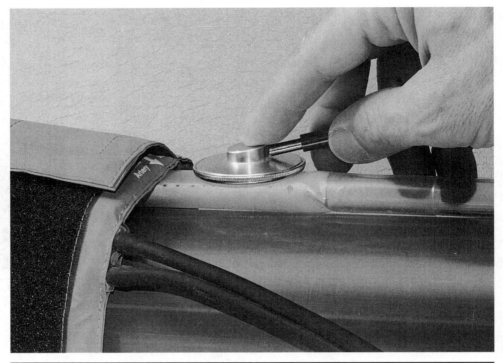

FIGURE 7-9   Measuring pressure in the tubing system with a blood pressure cuff and stethoscope.

tubing, just downstream from the cuff, slowly release the air from the cuff (Figure 7-9). You should hear the faint sound of fluid turbulence followed by loud "whooshing" and then silence as pressure diminishes.

Sounds were just noticeable on our system at about 160 mmHg and disappeared around 40 mmHg. If these readings were taken from a human, the systolic pressure of 160 mmHg would be acceptable, but with a diastolic pressure of only 40 mmHg, the person would probably suffer brain and other organ damage.

You can increase the equivalent of diastolic pressure to a more physiologically acceptable value by increasing the value assigned to `diastolicSpeed`. However, if you do so, the window of sound appearance to disappearance will narrow. As an extreme example, if `diastolicSpeed` and `systolicSpeed` are equal, then there is no pulse (ac component), but a constant flow (dc component) of water through the tubing.

Modifications

The system just described, as with most of the projects in this book, is intended to provide a basic, affordable experimental platform. If you plan to deploy the cardiac pulse simulator on an android or robot, then you'll need to modify the hardware accordingly. For example, you'll need to move from an open, large-volume system to a closed system with the minimum amount of fluid necessary for operation. A gallon or half-gallon paint can makes a great containment system for the pump. Paint cans are easy to drill, and brass or PVC fittings are readily available at hardware stores and online.

You might also consider upgrading the pump to something more compact, quieter, and capable of producing greater pressure or *head*. This generally means moving from plastic to metal housings and brass fittings. While you're at it, consider using a nonconductive fluid sold for PC cooling systems instead of water.

The modifications detailed herein assume that you're working with the basic platform described earlier. However, feel free to experiment with alternative or additional hardware. For example, liquid flowmeters are inexpensive and useful to determine exactly how much fluid is flowing through the system at various duty-cycle settings.

Peripheral Pulse Simulator

The cardiac pulse simulator creates the equivalent of an aorta, or at least one of the large femoral arteries in the legs. It's easy for even an inexperienced person to feel the pulse and hear Korotkoff sounds from such a large vessel. However, if you want to create a system with the tactile feel of pulses in the periphery, then you'll want to use ¼-in. tubing for the active palpation site.

We found that regular ¼-in. PVC tubing and silicone airline tubing are too stiff for the task. It simply takes too much fingertip pressure to partially occlude the lumen. However, an 8-in. length of thin-walled shrink-wrap tubing works great. An alternative to the shrink-wrap tubing is to use an 8-in. length of ¼-in. ID supersoft latex rubber tubing, available from McMaster-Carr.

Use a plastic tubing adapter to connect the ½-in. ID PVC tubing to the ¼-in. ID tubing and back again to the large PVC tubing for the return, as shown in Figure 7-10. Tie wraps will keep the shrink-wrap tubing in place for quick tests. For a more permanent connection, use silicone adhesive.

Because of the increased resistance of the smaller tubing and the adapters, the pulse that can be felt from the smaller tubing is significantly weaker than the pulse in the large tubing. However, if you're used to feeling for pulses, you'll appreciate the greater realism available through use of the smaller tubing. Similarly, you won't get the roaring Korotkoff sounds that you heard with the blood pressure cuff over the large tubing, but you should still hear the sounds clearly.

If you have trouble feeling the pulse, try developing your fingertip skills on your radial artery. Failing that, go for your carotids. If that fails, get a small hand mirror and hold it under your nose for a few seconds. If it doesn't fog, then you'll have an explanation for the lack of pulses.

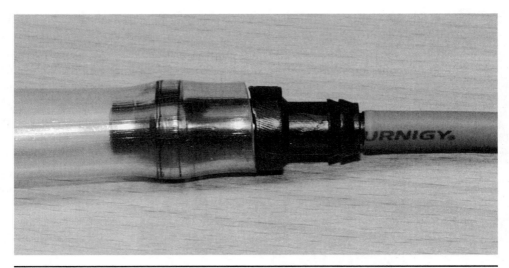

Figure 7-10 Close-up of ½-in. ID PVC tubing (*left*) interfaced with ¼-in. ID shrink-wrap tubing (*right*) with a tubing adapter.

Pulse Simulator with State Response

In this modification, we'll alter the pulse rate as a function of the status of a three-position switch. As with the examples in Chapter 6, the goal is to mimic physiology of someone sedated, sitting calmly, or exercising. Following the heart rates available through the heart sounds simulator, we'll use rates of 48, 80, and 108 beats/min for bradycardia, normal rate, and tachycardia, respectively.

Bill of Materials

In addition to the cardiac pulse simulator, you'll need either a pair of momentary contact switches or, preferably, a three-position toggle or sliding switch. Alternatively, just about any other sensor with simple on-off digital output can be used with the system without recoding. And analog sensors can be used instead of the switch with only a few added lines of code.

Circuit

The circuit, shown in Figure 7-11, consists of the original cardiac pulse simulator with the addition of a three-position switch, S1. If you opt to replace S1 with a pair of switches or sensors, then consider the condition in which both inputs are active simultaneously. You'll have to make provision for the condition in code.

Construction

Unless you opt for more complex state sensors, no additional construction is necessary.

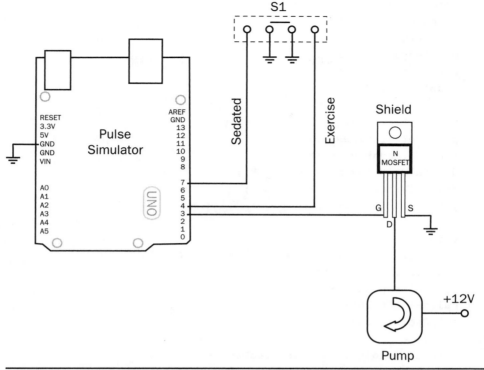

FIGURE 7-11   Cardiac pulse simulator circuit with three-position state switch S1.

Code

The Arduino code for the modified cardiac pulse simulator with responses for normal, sedated, and exercise states is shown in Listing 7-2.

LISTING 7-2   Arduino code for the cardiac pulse simulator.

```
/*
Cardiac Pulse Simulator With State Response
Androids: Build Your Own Lifelike Robots by Bergeron and Talbot
Listing 7-2. See www.mhprofessional.com/Androids for fully
documented code
Arduino 1.0.1 environment
*/

constintpumpPin = 3;
constintexercisePin = 4;
constintsedatedPin = 7;

intsystolicSpeed = 255;
```

```
intdiastolicSpeed = 100;
intpulseRate = 60;
intdutyCycle = 30;
doublesystolicDelay;
doublediastolicDelay;
doubletotalMS;

void setup() {
pinMode(exercisePin,INPUT);
pinMode(sedatedPin,INPUT);
digitalWrite(exercisePin,HIGH);
digitalWrite(sedatedPin,HIGH);
}

void loop() {
if (digitalRead (exercisePin) == LOW) {
systolicSpeed = 255;
diastolicSpeed = 80;
pulseRate = 108;
dutyCycle = 30;
    }
else if (digitalRead (sedatedPin) == LOW) {
systolicSpeed = 255;
diastolicSpeed = 120;
pulseRate = 48;
dutyCycle = 30;
    }
else {
systolicSpeed = 255;
diastolicSpeed = 100;
pulseRate = 80;
dutyCycle = 30;
    }
totalMS = (60000/pulseRate);
systolicDelay = (dutyCycle * totalMS)/100;
diastolicDelay = totalMS - systolicDelay;
analogWrite(pumpPin,systolicSpeed);
delay (systolicDelay);
analogWrite(pumpPin,diastolicSpeed);
delay (diastolicDelay);
}
```

The first thing to note is that we've added constants `exercisePin` and `sedatdPin` for pins D4 and D7. If you use the Sparkfun motor shield, pins D3, D5, D6, D9, D10, and D11 are unavailable for S1 or other connections. During `setup()`, these pins are defined as inputs, and the internal pull-up resistors are activated with the `digitalWrite()` function.

In the main loop, the two input ports are evaluated and the variables specifying the PWM waveform are defined depending on whether `exercisePin` or `sedatedPin` is `LOW`. The default values, when neither `exercisePin` nor `sedatedPin` is `LOW`, are a normal pulse rate and duty cycle of 30 beats/min.

Note that the values assigned to both `systolicSpeed` and `dutyCycle` are the same in all three cases. They're included in the listing so that you can tune your particular setup so that pulses are more or less prominent. If you decide to keep these values, then you can define these variables as constants at the beginning of the program. The values assigned to `diastolicSpeed` are different in each case to compensate for the inertia of the water in the system. You may need to experiment with different values depending on the configuration of your system.

Once values are assigned to the variables, the delay times between systole and diastole are calculated. The `analogWrite()` function is then used to vary the pump speed in the range defined by `systolicSpeed` and `diastolicSpeed`.

Operation

The setup and basic operation of the system is unchanged from that of the original cardiac pulse simulator. The only change is flipping S1 to select normal, slow, or rapid pulse. Because of the limitations of the pump, you may find the pulse weaker than with the original setting of 60 beats/min. At the slowest rate of 48 beats/min, the flow of water through the pump may come to a standstill before the next systolic pulse. That pulse has to overcome the inertia of the column of water along the length of the tubing. The work-around is to increase the value of `distolicSpeed` to keep the water moving.

You can also reduce the inertial dampening due to the water in the return leg of the system by shortening the tubing to the minimum length needed to reach the dishpan. Another step is to make certain that every part of the tubing is at or below the level of the pump. If you want to absolutely minimize the drag from gravity, then insert PVC connectors at both ends of the dishpan, at or below pump height, with one for input and one for output.

At the highest pulse rate of 108 beats/min, you'll probably find the pulse more difficult to detect, again because of limitations in the pump. In this case, the problem is due to a smaller difference between systolic and diastolic pressures. The parallel situation in electronics is the relative ease of smoothing a higher-frequency pulse train with a given capacitor. It's the reason a full-wave rectifier requires a smaller filter capacitor than a half-wave rectifier for the same dc output. One solution is to decrease the value assigned to `diastolicSpeed`, thereby increasing the difference between the diastolic and systolic pressures.

Integration with Heart and Breath Sound Simulators

Now that we have a cardiac pulse simulator with pulse rates that match the sound rates produced by the heart sound simulator from Chapter 6, combining the systems requires little more than adding details to the schematic in Figure 6-11.

Bill of Materials

At a minimum, you'll need the heart sound simulator we constructed in Chapter 6. Ideally, you'll have both the lung sound and heart sound simulators available. Add the cardiac pulse simulator with state response that we just finished and a few feet of wire, and you're done.

Circuit

The circuit, shown in Figure 7-12, is an embellishment of Figure 6-11. Recall that bringing pin A4 LOW on both the lung and heart sound simulators signals a sedated state accompanied by bradycardia and bradypnea. Bringing pin A5 low on both simulators signals exercise, with tachycardia and tachypnea. On the pulse generator, digital pins 8 and 12 are the sedation and exercise signaling pins, respectively.

As noted earlier, feel free to substitute a sensor or other input for S1. For example, you could base the state of the overall system on the state of the tilt sensor in the heart sound simulator.

The three simulators are loosely integrated, and that's good enough for our purposes. Given that the MP3 files of heart sounds all start with the "lub" of contraction, the pulsations created by the pump will sync with the sounds associated with the valves.

You may be wondering why we don't simply connect switch S1 in parallel with the control pins of the heart and lungs sound simulators. Programmatic control of the simulators enables the system to simulate a wider range of clinical possibilities. For example, in some conditions, such as a stab wound or advanced atherosclerosis, there simply isn't a palpable pulse despite contraction of the heart and audible heart sounds.

Construction

The only construction detail of note is to take provisions to prevent a gusher of water from frying your Arduino boards. Simply inserting your boards into plastic freezer bags should cover most accidents. Liberal use of gaffer's tape can also keep tubes and wires from moving unexpectedly.

Code

The Arduino code for the integrated cardiac pulse simulator is shown in Listing 7-3. The code is intended to work with the heart and lung sound simulator programs in Listings 6-5 and 6-6.

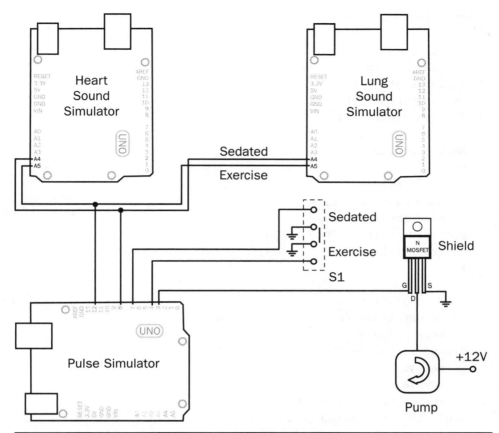

FIGURE 7-12 Schematic of cardiac pulse simulator integrated with heart sound and lung sound simulators.

LISTING 7-3 Arduino code for integrated cardiac pulse simulator.

```
/*
Integrated Cardiac Pulse Simulator
Androids: Build Your Own Lifelike Robots by Bergeron and Talbot
Listing 7-3. See www.mhprofessional.com/Androids for fully
documented code
Arduino 1.0.1 environment
*/

constintpumpPin = 3;
constintexercisePin = 4;
constintsedatedPin = 7;
constintexerciseOutPin = 8;
```

```
constintsedatedOutPin = 12;

intsystolicSpeed = 255;
intdiastolicSpeed = 100;
intpulseRate = 60;
intdutyCycle = 30;
doublesystolicDelay;
doublediastolicDelay;
doubletotalMS;

void setup() {
pinMode(exercisePin,INPUT);
pinMode(sedatedPin,INPUT);
digitalWrite(exercisePin,HIGH);
digitalWrite(sedatedPin,HIGH);
}

void loop() {
if (digitalRead (exercisePin) == LOW) {
systolicSpeed = 255;
diastolicSpeed = 80;
pulseRate = 108;
dutyCycle = 30;
digitalWrite(exerciseOutPin,LOW);
digitalWrite(sedatedOutPin,HIGH);
   }
else if (digitalRead (sedatedPin) == LOW) {
systolicSpeed = 255;
diastolicSpeed = 120;
pulseRate = 48;
dutyCycle = 30;
digitalWrite(exerciseOutPin,HIGH);
digitalWrite(sedatedOutPin,LOW);
   }
else {
systolicSpeed = 255;
diastolicSpeed = 100;
pulseRate = 80;
dutyCycle = 30;
digitalWrite(exerciseOutPin,HIGH);
digitalWrite(sedatedOutPin,HIGH);
}
```

```
totalMS = (60000/pulseRate);
Serial.println (totalMS);
systolicDelay = (dutyCycle * totalMS)/100;
diastolicDelay = totalMS - systolicDelay;
analogWrite(pumpPin,systolicSpeed);
delay (systolicDelay);
analogWrite(pumpPin,diastolicSpeed);
delay (diastolicDelay);
}
```

Compared with the previous listing for the cardiac pulse simulator, there are two new constants, `exerciseOutPin` and `sedatedOutPin`, which are assigned digital pins 8 and 12, respectively. When the pin linked to `ExerciseOutPin` is brought LOW, pin A4 on both heart and lung sound simulators is brought LOW, causing the simulators to generate sounds characteristic of an exercising state—tachycardia and tachypnea. Conversely, when `sedatedOutPin` is brought LOW, pin A5 on both sound simulators is brought LOW, causing them to generate sounds characteristic of a sedated state—bradycardia and bradypnea.

The digital pins associated with `exerciseOutPin` and `sedatedOutPin` are assigned HIGH or LOW values depending on the evaluation of the pins associated with S1, `sedatedPin` and `exercisePin`. The default, both `sedatedPin` and `exercisePin` HIGH, is mirrored by the values of `exerciseOutPin` and `sedatedOutPin`. As a result, the heart and lung sound simulators produce normal sounds that are accompanied by a normal pulse rate from the pulse simulator.

Operation

To appreciate the integration of the simulators, you'll want to listen to the heart and then the lung sounds while palpating the cardiac pulse. As with a real person, doing so can be a challenge. You need one hand to hold and operate the stethoscope and the other to palpate the vessel. However, if you arrange your workspace properly, you should be able to appreciate the integration of physical pulses and audio as you use S1 to change the scenario from resting to sedated to exercising.

Gremlins

The greatest challenge with this project was identifying affordable, easy-to-work-with parts capable of demonstrating cardiac pulses. With a $250 budget, it's simply a matter of visiting a professional supply site, such as McMaster-Carr, and ordering a powerful all-metal pump and all the necessary fittings and tubing. Recognizing that this option isn't available to most readers, we've done our best to keep the overall project affordable, simple, and repurposable. We picked up the clear PVC tubing and adapters at a local ACE Hardware store and the pump from Amazon.

Also, it's not like we just walked into a hardware store and walked out with the solution. In developing the materials for this chapter, we also created a complete system using inexpensive ½-in. garden hose, as well as a complete system based on aquarium airline. The numerous valves, fittings, and splitters available in garden and pet-supply shops provide considerable flexibility. For example, with a simple Y hose adapter or aquarium gang valve, it's a simple matter to adjust the relative flow of water through two parallel lines—one representing the flow to the skeletal muscles and one to the internal organs.

If you're going for maximum realism—say, you're developing teaching mannequins for nursing or medical students or a simulated dog lab—then you'll want to use tubing with the same diameter and "feel" of real arteries. Check out the McMaster-Carr online catalog, which has a special tubing-finder utility. In particular, take a look at the supersoft latex rubber tubing we mentioned earlier. We developed a pulse simulator using this tubing, in various diameters, from the pump to the limb-sized waterproof cylinder. It worked too well. The tubing absorbed the pulses from the pump to such an extent that there was barely a detectable pulse only a few feet from the pump. The solution was a more powerful—and considerably more expensive—pump.

The bottom line is not to feel constrained by the particular choices we've made in the demonstration platforms described here. There are multiple, affordable options, as well as high-end solutions, depending on your needs.

Search Terms

Try the following search terms for your browser:

- Korotkoff sounds
- Pedal pulses
- Radial artery
- Dorsalispedis artery
- Femoral artery
- Brachial artery
- Palpation
- Water pump head
- mmHg

CHAPTER 8

Simply a
Matter of Time

Ever notice how otherwise perfect androids are often uncovered because they never age? Science fiction authors have long contemplated limiting the lifespan of androids for the good of humankind—take the replicants in *Blade Runner* that were engineered to live only four years versus the Cylons in *Battle Star Galactica* that are initially immortal. Most authors conclude that whereas individual autonomous beings naturally desire to live forever, such a condition is incompatible with humanity. Our resources and autonomy would eventually be usurped by our creations, which would self-replicate into increasingly superior beings.

Considerations of intentionally limiting the function and longevity of an android may seem preposterous at this stage of robotics evolution. Think about it, we celebrate when a Mars rover lasts several months past the planned failure date, and we strive to build robots that can survive a few accidental crashes, much less outlive us.

However, there are instances when forced death or destruction of our mechatronic creations is warranted. Consider the aberrant autonomous missile that veers off course and must be terminated with an autodestruct instruction. And you may have firsthand experience with car and electronics manufacturers that design their products to self-destruct a few days after the warranty expires in order to maximize profits.

Illness and age go hand in hand. In fact, some people believe that aging is a disease that should be eradicated. That's probably a few generations away. It's more likely that we'll devise machines that outlive us first. With this in mind, we'll explore how human biological systems age and fail and how to mimic the appropriate physiology and behavior. Get comfortable behind your keyboard because most of the experiments involve software upgrades to modifications of previously described projects, spiced up with a few sensors and feedback devices where appropriate.

Muscular Strength and Aging

As we age, our external body parts generally become bigger, hairier, and lower to the ground. We also tend to lose muscle mass and strength. And if all that weren't bad enough, there are diseases that accelerate the mental and physical aging processes. Given the complexity of modeling cognitive decline, let's consider the more straightforward physical and behavioral aspects of aging and disease.

Biological Basis

One of the diseases that prematurely robs humans of their muscular strength is myasthenia gravis ("grave muscle weakness"), which is caused by a breakdown of the communications between nerves and the voluntary skeletal muscles. The result is both weakness and rapid fatigue. As illustrated in Figure 8-1, myasthenia gravis (MG) is associated with abnormal neuromuscular junctions—specifically, the receptors are attacked by a person's own antibodies, leaving fewer sensors to respond to impulses sent down a motor fiber.

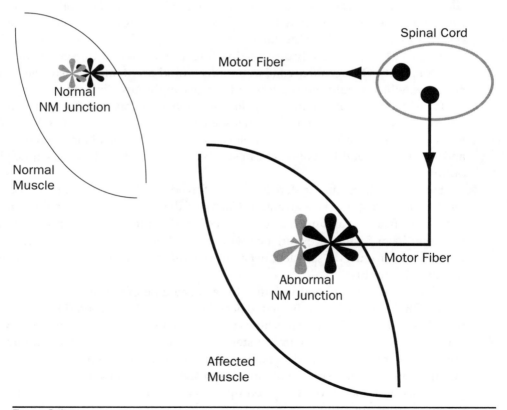

FIGURE 8-1 Myasthenia gravis is associated with defective receptors on the muscle side of the neuromuscular (NM) junction.

MG is the most common disorder affecting the neuromuscular junction. The disease is especially problematic because the muscles affected are the ones that are used repeatedly—including muscles that move the eyes and eyelids and that are involved in chewing and swallowing. Moreover, the disease progresses with time.

MG is often confused with muscular sclerosis (MS). Although both result in muscular weakness, MS is due to failure of the fibers that carry signals from sensors and the brain to the muscle fibers. The neuromuscular junction is not affected in MS.

Figure 8-2 shows the responses of a normal muscle and a muscle affected by MG to repeated, rapid stimulation. Whereas the strength of contraction is constant in the normal muscle, the affected muscle fatigues within a few seconds. As you might gather from the figure, the disease does not affect the results of a standard reflex test. A single strike with a reflex hammer to a muscle tendon at the knee or elbow to elicit a reflex response doesn't fatigue the muscle.

In normal muscle tissue, there is an all-or-none response to a neural stimulus. Although the refractory period affects when the muscle will respond to subsequent

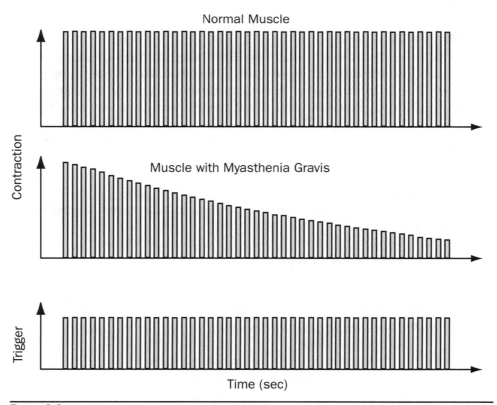

FIGURE 8-2   Normal muscle response (*above*) versus muscle affected by MG (*center*).

stimulation, when it does respond, it does so fully, as in the top tracing in the figure. With MG, there is also a refractory period, but the muscle response is progressively diminished, as in the middle tracing. With rest, the neuromuscular junction recuperates.

In the examples shown in Figure 8-2, frequency of stimulation is low enough that the refractory period we discussed in Chapter 1 is not a factor. The differences in the response patterns are due to the disease.

Relevance to Android Designs

The ability to mimic fatigue is useful in a variety of contexts. The first is bonding with human coworkers. A soldier android can show signs of fatigue when appropriate and then drop the acting and kick it into high gear when necessary.

There's also the current, practical application of androids used as patients for training military medics and clinicians. For example, there are many causes of fatigue and muscle weakness, from heart disease and poor nutrition to the effects of shrapnel. Enabling an android to mimic a variety of disease presentations ultimately helps clinicians to learn to recognize and treat a variety of disorders and diseases.

Experiment

In this experiment, we'll model the fatigability of normal, healthy muscle and diseased skeletal muscle with repeated stimulation. We'll ignore the complexity of a software refractory period in this experiment because we're using a robot jaw with an inherent mechanical refractory period. Also, we ideally would like a robot gripper that could open and close several times a second, but the cost of the associated hardware is prohibitive. Instead, we'll repurpose materials from previous experiments.

Bill of Materials

In addition to an Arduino Uno or equivalent and associated hookup wire, you'll need the following:

- Robot gripper (arm optional)
- Force-sensing resistor
- 10-kΩ resistor
- Small 1/8-in.-thick adhesive rubber foot
- Momentary contact button switch
- 10-kΩ potentiometer
- 5-V direct-current (dc) power supply for servo
- Shield of your choice (optional)

If you built one of the systems described in Chapter 2, then you already have everything you need. We're using a Lynxmotion AL5B robot arm gripper with a small Sparkfun 100-g to 10-kg force-sensing resistor. A Grove shield, momentary contact switch, and 10-kΩ potentiometer make for quick assembly.

Circuit

The circuit for the MG simulator is shown in Figure 8-3. Potentiometer P1, the severity control, provides a 0- to 5-V input to pin A1. Momentary contact switch S1 grounds digital pin 5, signaling the gripper jaws to close. The force-sensing resistor FSR1 and 10-kΩ resistor R1 form a voltage divider circuit.

With no force applied, the resistance of FSR1 is several orders of magnitude greater than that of R1. As a result, most of the voltage drop occurs across FSR1, with only a drop of a few millivolts across R1. With increasing pressure on FSR1 and resulting lower resistance, the voltage across R1 increases. Note the separate 5-V power connection to the gripper servo. If you try to power the servo from the Arduino's 5-V bus, you risk frying the onboard voltage regulator.

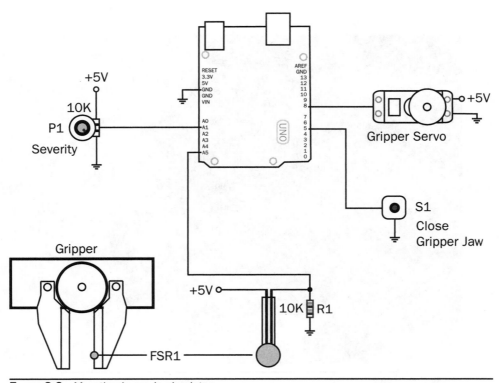

FIGURE 8-3 Myasthenia gravis simulator.

Construction

Assuming that you have a gripper assembled, the only real construction is attaching the force-sensing resistor to one jaw of the gripper assembly, as we did in Chapter 2. If you use the Sparkfun sensor, then remove the film covering the adhesive backing, and attach the sensor directly to the hard plastic jaw. Next, attach the self-adhesive rubber foot directly on the face of the force-sensing resistor, as in Figure 8-4. The foot distributes force evenly on the sensor and protects the surface from puncture by hard objects. Also, feel free to use a different input and output pin mapping, just as long as you modify the code accordingly.

Code

The Arduino code for the MG simulator is shown in Listing 8-1.

LISTING 8-1   Arduino code for myasthenia gravis simulator.

```
/*
Myasthenia Gravis Simulator
Androids: Build Your Own Lifelike Robots by Bergeron and Talbot
Listing 8-1. See www.mhprofessional.com/Androids for documented
code
```

FIGURE 8-4   Force-sensing resistor and gripper assembly.

```
Arduino 1.0.1 environment
Gripper strength rapidly declines with repeated use;
Recovers with rest
*/

#include <Servo.h>
Servo gripperServo;
const int FSRPin = A5;
const int P1Pin = A1;
const int S1Pin = 5;
const int gripperServopin = 8;
const int servoSettleDelay = 30;
const int sensorRange = 810;
const long minimumRestMillis = 10000;

int FSRValue = 0;
int FSROffset = 0;
int P1Value = 0;
int S1ButtonState = 1;
int grippperServoPosition = 0;
unsigned long currentMillis = millis();
unsigned long lastMillis = millis();

void setup() {
  gripperServo.attach(gripperServopin);
  pinMode(S1Pin,INPUT);
  digitalWrite(S1Pin,HIGH);
  openJaws();
  reset();
}

void loop(){
  readSeverity();
  S1ButtonState = digitalRead(S1Pin);
    if (S1ButtonState == LOW) {
      Serial.println(P1Value);
      currentMillis = millis();
      if (currentMillis - lastMillis > minimumRestMillis) {
        reset();
      }
        else {
        lastMillis = currentMillis;
```

```
        }
     FSROffset = FSROffset + P1Value;
     if(FSROffset >= sensorRange){
       FSROffset = sensorRange - P1Value;
     }
     openJaws();
     delay (300);
     closeJaws();
     delay (600);
     openJaws();
     }
  }
  /*
  ------------------------------------------------------------
  readSeverity()
  openJaws()
  closeJaws()
  reset()
  ------------------------------------------------------------
  */

  void readSeverity() {
  P1Value = analogRead(P1Pin);
  P1Value = map(P1Value, 0, 1023, 0, 100);
}

void openJaws(){
    grippperServoPosition = 0;
    gripperServo.write(grippperServoPosition);
}

void closeJaws(){
  for (grippperServoPosition = 0; grippperServoPosition < 180;
grippperServoPosition++) {
FSRValue = analogRead(FSRPin);
        if (FSRValue < sensorRange - FSROffset) {
      gripperServo.write(grippperServoPosition);
      delay (servoSettleDelay);
    }
  }
}
```

```
void reset() {
   lastMillis = currentMillis;
   FSROffset = 0;
}
```

Starting at the top, the constants for the sensors and controls are defined. Note the values for servoSettleDelay, sensorRange, and minimumRestMillis because you'll probably need to change these values to fit your particular hardware environment. Even if you use the suggested hardware, you'll probably need to use a different value for sensorRange simply because of normal manufacturing variation in sensors, servos, and grippers.

To this end, not shown here but available in the online version of the listing are multiple print statements that enable you to trace the value returned by the force-sensing resistor over repeated cycles. Alternatively, you can run the short calibration program that is described and listed below. Substitute the maximum value returned by the program for the value of sensorRange.

Returning to Listing 8-1, of note in the variable declaration area are the variables for current time, currentMillis, and for the time since the last jaw closure, lastMillis. Checking the elapsed time is more efficient than using the delay function because this doesn't block other events and obviates the need to use interrupts.

During setup, the jaws are opened, and the elapsed time clock is set to zero. The force-sensing resistor offset, FSROffset, is also set to zero. The offset is used to incrementally decrease the force applied by the gripper claw.

In the main loop, the readSeverity() function reads the value of P1 and scales the 10-bit reading from the 0 to 5 input to 0–100. The scaled result, stored in P1Value, represents normal (*P1Value* = 0) to severe MG (*P1Value* = 100). When S1 is pressed, the time since the last button press is calculated, and if the minimum rest time, minimumRestMillis, has passed, then the clock is reset.

If, however, the minimum rest period has not passed, then FSROffset is incremented by P1Value. When closeJaws() is called, the for loop increments gripperServoPostion to full clockwise rotation. As the jaws of the gripper increase pressure on the force-sensing resistor, the value assigned to FSRValue increases. As long as the value of FSRValue is less than sensorRange – FSROffset, the updated value of gripperServoPostion is written to the servo. The delay call allows the servo to settle before the next increment in gripperServoPostion is sent to the servo.

Following, in Listing 8-2, is the force-sensor calibration utility. The program assumes the current hardware configuration with the same pin assignments. When you run the utility, make certain that the full force of the jaw is applied to the rubber foot covering the force resistor. Depending on the design of your robot jaw, you may have to put a hard object in the jaw to ensure that the full force is applied to the force-sensing resistor.

LISTING 8-2   Arduino code for determining the maximum value returned by the force-sensing resistor.

```
/*
Force Sensor Calibration Utility
Androids: Build Your Own Lifelike Robots by Bergeron and Talbot
Listing 8-2.
Arduino 1.0.1 environment
*/

#include <Servo.h>
Servo gripperServo;
const int FSRPin = A5;
const int gripperServopin = 8;
const int servoSettleDelay = 30;
int FSRValue = 0;
int sensorRange = 0;
int grippperServoPosition = 0;

void setup() {
  Serial.begin(9600);
  gripperServo.attach(gripperServopin);
}

void loop(){
    if (FSRValue == 0) {
     openJaws();
     delay (300);
     closeJaws();
     delay (600);
     openJaws();
     Serial.print("sensorRange = ");
     Serial.println(sensorRange);
    }
}

  /*
  ------------------------------------------------------------
  openJaws()
  closeJaws()
```

```
--------------------------------------------------------------
*/

void openJaws(){
    grippperServoPosition = 0;
    gripperServo.write(grippperServoPosition);
}

void closeJaws(){
for (grippperServoPosition = 0; grippperServoPosition < 180;
grippperServoPosition++) {
        gripperServo.write(grippperServoPosition);
        FSRValue = analogRead(FSRPin);
        if (FSRValue > sensorRange){
          sensorRange = FSRValue;
        }
        delay (servoSettleDelay);
    }
  }
```

Most of the action is around the `closeJaws()` function. As the servo and attached jaws are incrementally swept from open to closed position, the force-sensing resistor is read. If the current reading of the force-sensing resistor is greater than the value of `sensorRange`, then the value of `sensorRange` is updated to equal the current reading. At the end of the sweep, the maximum value read from the force-sensing resistor during the run is output to the serial monitor.

Operation

Assuming that the you ran the utility program and updated `sensorRange` to reflect your setup, the next step is to put either your thumb (nail toward the force-sensing resistor) or a soft cube of foam in the jaws. With P1 rotated fully clockwise, press S1. The jaws will close, initially to the full strength of the servo or the full range of the force-sensing resistor, whichever is less. Unless you have a higher-end robot arm or gripper, the servo will probably top out before the force sensor. Our standard Lynxmotion AL5B robot arm gripper with diminutive HS-645MG servo and Sparkfun 100-g to 10-kg force-sensing resistor tops out at a sensor reading of about 810 (3.97V).

After the squeeze, the jaws will open automatically, awaiting your next button press. If you press S1 before the minimum recovery time—10 seconds in the example here—then the force of the next closure will be equal to or less than the pressure during the previous closure. You should be able to feel the decrease in pressure with each closure of the jaws until the system bottoms out at a modest closure pressure that's significantly less than the pressure at the beginning of the sequence.

Now vary the position of potentiometer P1 to modify the severity of MG. Fully counterclockwise, the gripper will respond normally, with repeatable pressure from one closure to the next. As you rotate P1 clockwise, the rate of fatigue increases. With P1 rotated fully clockwise, closure pressure rapidly decreases and bottoms out after about six cycles. With P1 rotated to the midpoint, the decrease in pressure bottoms out after about a dozen cycles.

If you're using this setup as a demonstration for a group, then a soft cube of foam makes a better target than your thumb because the degree of compression is visible (Figure 8-5). The photo on the left of the figure shows end compression after the first button press. The photo on the right shows end compression after six button presses with P1 fully clockwise. There is obviously more compression at the start of the cycle than at the end.

If you don't see similar differences in compression with progression of the cycle, then use softer foam. We found the pick-and-pluck foam that accompanies Pelican and SKB cases perfect for the little Lynxmotion gripper. Whatever foam you use, position it so that it primarily contacts the force sensor and not the surrounding jaw of the gripper. Otherwise, the force sensor reading will be artificially depressed.

Modifications

One of the limitations of standard servos and grippers is the small difference between no force and full-scale force when the system is used with a hard,

FIGURE 8-5   Full compression after one cycle (*left*) and after six cycles (*right*) with potentiometer P1 fully clockwise.

noncompressible object. For a ballpoint pen, for example, one increment in `gripperServoPosition` results in a pressure-reading jump of about 500 (2.45V).

If you need a system that's compatible with hard objects, then you should consider constructing your own gripper or compressor with a gearbox or servo reducer. For example, if you use a 5:1 reducer, then that same 500 (2.45V) jump in sensor readings would be spread out over five increments, or 100 (490mV) per increment.

If money is no object, then commercial heavy-duty servo reducers with ratios of up to 3,300:1 are readily available. However, if you don't need to crush rocks and have a modest budget, then consider a ServoCity aluminum and stainless steel power gearbox with a 7:1 gear ratio. At around $100 plus the price of the servo, it's a considerable investment. An intermediate solution is to use a linear actuator with a multiturn worm gear for a drive system.

On the other hand, if you need to work with even softer or more fragile materials, then consider using force-sensitive resistors with greater sensitivity. A minor downside is that many of the high-sensitivity force-sensitive resistors require an op amp between the sensor and the Arduino's analog input.

Another modification worth considering is to use the onboard light-emitting diode (LED), an external LED, or a buzzer to signal when the system has idled long enough for full strength recovery. Or you could connect the Arduino to a laptop running Processing and create an interactive bar graph showing instantaneous and past force readings.

Heart Recovery Rate and Fitness Level

If there's one thing that distinguishes the young from the old and the active from the sedentary, it's cardiovascular fitness. Although there are exceptions, older humans in this culture tend to spend more time sitting than sprinting. You could argue over the aesthetics of someone 35 pounds overweight, but from a cardiovascular perspective, there's no getting away from the truth. Heart recovery rate—the time it takes someone's heart rate to return to normal after an intense period of exercise—is a key indicator of cardiovascular fitness and of mortality.

Biological Basis

As we discussed in Chapter 7, maximum heart rate is generally limited by age. Someone age 35 should have a maximum heart rate of around $220 - 35 = 185$ beats/min. And if that person is fit, he'd typically exercise at up to 80 to 85 percent of his maximum. In the case of a 35-year-old, that's about 150 beats/min. More important, the fit person's heart will quickly recover, dropping the rate down to preexercise levels in a few minutes.

The heart of a cubicle rat or couch potato, in contrast, might tack away at an elevated rate for 10 or 20 minutes following a bout of intense exercise and require a

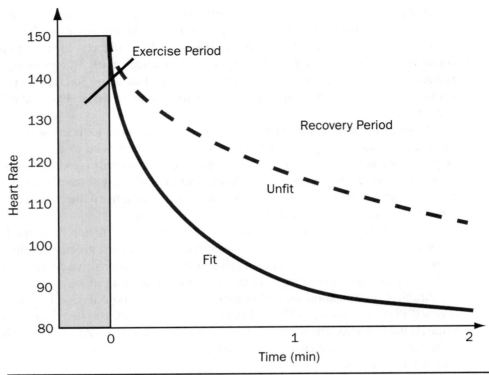

FIGURE 8-6   Typical heart rate recovery curves associated with different levels of cardiovascular fitness.

day or more to return to the normal resting rate. And if someone is really out of shape, he has a fair chance of never making it to the recovery period compared with someone who exercises regularly—young or old, a sedentary person is 50 times more likely to suffer a heart attack during intense exercise.

The rate of heart rate recovery versus level of cardiovascular fitness is illustrated in Figure 8-6. Assume that we have two 35-year-old subjects, one a regular runner ("fit") and one a couch potato ("unfit"). Even though it's a stretch, to keep this example simple, let's say that they both have a resting pulse rate of 80 beats/min. After exercising at 80 percent maximum heart rate, both rest for two minutes. The fit person's heart rate will drop precipitously in the first minute and then relatively slowly thereafter. By two minutes, the heart rate will be slightly elevated over the normal resting rate. In contrast, the unfit person's heart rate will slowly decrease, asymptotically, toward the preexercise resting pulse.

Relevance to Android Designs

As noted earlier, the more accurately an android mimics humans, the closer the potential bond. In the simple simulator in Chapter 7, heart rate was elevated during exercise, but there was no provision for variable recovery time to reflect age and fitness level. The following experiment illustrates how the original code can be extended to provide an android with a more lifelike physiologic response to exercise and exertion.

Experiment

The experiment described here is a modification of the heart sound simulator presented in Chapter 6. The following discussion assumes that you at least read through Chapter 6.

The goal is to play heart sounds that are appropriate for the location on the chest wall and that reflect the cardiovascular physiology of two 35-year-old subjects—one fit and one unfit. Using Figure 8-6 as a guide, we'll create a simulator that produces heart sounds at 150 beats/min to simulate the exercise response. When triggered, the heart rate recovery will mimic that of either a fit or unfit individual depending on a switch setting. The postrecovery heart rate will also reflect the fitness of the subject.

The heart sound mapping to the supplied audio files for this experiment is as follows:

- `track001.mp3`: Normal (A)—80 beats/min
- `track002.mp3`: Normal (P)—80 beats/min
- `track003.mp3`: Normal (T)—80 beats/min
- `track004.mp3`: Normal (M)—80 beats/min
- `track009.mp3`: Elevated (A)—108 beats/min
- `track010.mp3`: Elevated (P)—108 beats/min
- `track011.mp3`: Elevated (T)—108 beats/min
- `track012.mp3`: Elevated (M)—108 beats/min
- `track024.mp3`: Exercising (A)—150 beats/min
- `track025.mp3`: Exercising (P)—150 beats/min
- `track026.mp3`: Exercising (T)—150 beats/min
- `track027.mp3`: Exercising (M)—150 beats/min
- `track028.mp3`: Fit recovery (A)—150–82 beats/min
- `track029.mp3`: Fit recovery (P)—150–82 beats/min
- `track030.mp3`: Fit recovery (T)—150–82 beats/min
- `track031.mp3`: Fit recovery (M)—150–82 beats/min
- `track032.mp3`: Unfit recovery (A)—150–106 beats/min
- `track033.mp3`: Unfit recovery (P)—150–106 beats/min
- `track034.mp3`: Unfit recovery (T)—150–106 beats/min
- `track035.mp3`: Unfit recovery (M)—150–106 beats/min

The approximate fitness mapping for heart rate during fit and unfit recovery is shown in Table 8-1. As shown in the table, the heart rate of the sounds associated

TABLE 8-1   Stereotypic Heart Rate Recovery Beats per Minute versus Time for 35-Year-Old Fit and Unfit Subjects

Time (s)	0	15	30	45	60	75	90	105	120
Fit (beats/min)	150	115	102	96	90	88	86	84	82
Unfit (beats/min)	150	134	128	124	118	112	110	108	108

with areas in which each of the four valve sounds are heard best starts at 150 beats/min and progresses rapidly to 82 beats/min in the fit case or slowly to 108 beats/min in the unfit case.

Bill of Materials

The main components of the heart rate recovery simulator are listed below. Note the two switches, a momentary contact switch and an SPST slider or rocker switch. There is no positional detection sensor in this experiment. If you're using the setup from Chapter 6, then temporarily disconnect the mercury tilt switch from the Arduino circuit.

- Arduino Uno or equivalent
- 5-Vdc power supply
- Sparkfun MP3 player shield
- 1-GB microSD, FAT 32, with MP3 sound files
- TPA2005D1 1.6-W audio amp or equivalent
- 15-kΩ, 1/8- or 1/4-W resistors (2)
- Sparkfun small surface transducer
- Optek Technology OH090U Hall effect sensors (4)
- Momentary contact switch, normally open
- SPST slider or rocker switch
- Thin, nonmetallic mounting surface (mannequin optional)
- Small rare-earth magnet
- Stethoscope with aluminum head (optional)

If you didn't build the circuit described in Chapter 6 and want to experience the simulator with a minimal investment in parts, then you can get by with the following:

- Arduino Uno or equivalent
- Sparkfun MP3 player shield

- 1-GB microSD, FAT 32, with MP3 sound files
- Headphones
- 10-kΩ resistors (4)
- Male-to-male jumper wires

Figure 8-7 shows the Sparkfun MP3 player shield with microSD card installed and a few of the male-to-male jumper wires. The headers, sold separately, will have to be installed on the shield before you can use the wires or install the shield on an Arduino. If you're new to the use of prefabricated jumpers, then consider adding an assortment of male-to-male, male-to-female, and female-to-female jumpers to your toolkit. They're inexpensive, great time savers, and available from all the major online vendors.

Circuit
The circuit for the heart rate recovery simulator, shown in Figure 8-8, is a minor modification of the circuit described in Chapter 6. The difference is that S1 is a momentary contact switch instead of a mercury tilt switch and S2 is an SPST slider or rocker switch.

Figure 8-7   Major components for a minimalist heart rate recovery simulator.

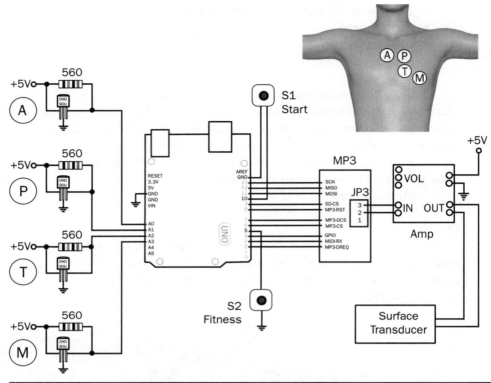

Figure 8-8 Circuit for heart rate recovery simulator. S2 is an SPST rocker/slider switch. S1 is a momentary contact switch.

As a review, the four Hall effect sensors are positioned in a board or mannequin to approximate the locations at which the aortic (A), pulmonic (P), tricuspid (T), and mitral (M) valve sounds are heard best. When a stethoscope with attached magnet is brought near a Hall effect sensor, the sensor conducts, bringing the associated analog port LOW.

The Sparkfun MP3 player board, TPA2005D1 1.6-W audio amplifier, and Sparkfun small surface transducer handle the playback of the appropriate MP3 files stored on the microSD card. As discussed in Chapter 6, feel free to replace the sensors and output electronics to suit your budget and needs.

If you opt for the minimalist approach, then plug your headphones into the MP3 player shield, use the four 10-kΩ resistors to pull up the analog input ports to 5 V, and use two of the wires in place of S1 and S2. Use a third wire to selectively bring one of the analog ports LOW to simulate an activated Hall effect sensor.

Construction

Construction follows the details given in Chapter 6. If you're using the circuit described in Chapter 6, then simply mount S1 and S2 where they're convenient, and disconnect the mercury tilt switch. If you're following the minimalistic approach, then mount the MP3 shield on the Arduino Uno, attach the USB cable for power, and keep the pigtail wires handy.

Code

LISTING 8-3 Arduino code for simulating heart rate recovery that mimics the physiologic response for fit and unfit individuals.

```
/*
Heart Rate Recovery Simulator
Androids: Build Your Own Lifelike Robots by Bergeron and Talbot
Listing 8-3. See www.mhprofessional.com/Androids for documented
code
Arduino 1.0.1 environment
Uses the Sparkfun MP3 Shield Library by Bill Porter
*/

#include <SPI.h>
#include <SdFat.h>
#include <SdFatUtil.h>
#include <SFEMP3Shield.h>

SFEMP3Shield MP3player;

const int fitPin = 5;
const int startPin = 10;
const int Aortic = 14;
const int Pulmonic = 15;
const int Tricuspid = 16;
const int Mitral = 17;

int soundTrack = 1;
int index = 0;

void setup() {
pinMode(fitPin,INPUT);
  pinMode(startPin,INPUT);
  pinMode(Aortic,INPUT);
  pinMode(Pulmonic,INPUT);
  pinMode(Tricuspid,INPUT);
  pinMode(Mitral,INPUT);
  digitalWrite(fitPin,HIGH);
```

```
    digitalWrite(startPin,HIGH);
    MP3player.begin();
    MP3player.SetVolume(1,1);
}

void loop() {
  readStartPin();
  if (index == 0) ExerciseRate();
  if (index == 1) RecoveryRate();
  if (index == 2) PostRecoveryRate();

if (!MP3player.isPlaying()){
if (index == 1) index = 2;
    MP3player.playTrack(soundTrack);
}
 delay(10);
}

 /*
 ------------------------------------------------------------
readStartPin()
ExerciseRate()
RecoveryRate()
PostRecoveryRate()
 ------------------------------------------------------------
 */

void readStartPin(){
    if (digitalRead (startPin) == LOW) {
    if (index == 0) index = 1;
    if (index == 2) {
      index = 0;
      MP3player.stopTrack();
    }
    delay (100);
}
}

void ExerciseRate(){
  if (digitalRead (Aortic) == LOW) soundTrack = 24;
  if (digitalRead (Pulmonic) == LOW) soundTrack = 25;
  if (digitalRead (Tricuspid) == LOW) soundTrack = 26;
```

```
    if (digitalRead (Mitral) == LOW) soundTrack = 27;
    if (soundTrack <24 || soundTrack >27) soundTrack = 24;
}


void RecoveryRate(){
  if (digitalRead (fitPin) == LOW){

  if (digitalRead (Aortic) == LOW) soundTrack = 28;
  if (digitalRead (Pulmonic) == LOW) soundTrack = 29;
  if (digitalRead (Tricuspid) == LOW) soundTrack = 30;
  if (digitalRead (Mitral) == LOW) soundTrack = 31;
  if (soundTrack <28 || soundTrack >31) soundTrack = 28;
  }
  else {
  if (digitalRead (Aortic) == LOW) soundTrack = 32;
  if (digitalRead (Pulmonic) == LOW) soundTrack = 33;
  if (digitalRead (Tricuspid) == LOW) soundTrack = 34;
  if (digitalRead (Mitral) == LOW) soundTrack = 35;
  if (soundTrack <32 || soundTrack >35) soundTrack = 32;
  }
}


void PostRecoveryRate(){
  if (digitalRead (fitPin) == LOW){

  if (digitalRead (Aortic) == LOW) soundTrack = 1;
  if (digitalRead (Pulmonic) == LOW) soundTrack = 2;
  if (digitalRead (Tricuspid) == LOW) soundTrack = 3;
  if (digitalRead (Mitral) == LOW) soundTrack = 4;
   if (soundTrack <1 || soundTrack >4) soundTrack = 1;
  }
  else {
  if (digitalRead (Aortic) == LOW) soundTrack = 9;
  if (digitalRead (Pulmonic) == LOW) soundTrack = 10;
  if (digitalRead (Tricuspid) == LOW) soundTrack = 11;
  if (digitalRead (Mitral) == LOW) soundTrack = 12;
  if (soundTrack <9 || soundTrack >12) soundTrack = 9;
  }
}
```

Because of the sequencing, the preceding code is significantly different from that described in Chapter 6. One thing that's remained constant is the excellent

Sparkfun MP3 shield library by Bill Porter. It's possible to achieve the same functionality without the library, but it'll cost you several hours of work.

Starting at the top, the important features are the constants defined for the fitness switch port, `fitPin`, and the starting switch port, `startPin`. The port-assignment options are limited because the Hall effect sensors and MP3 player card consume most of the digital and analog ports.

The `soundTrack` variable determines which MP3 file to play, where the track assignment n is translated to `track00n.mp3`. The variable `index` is used to switch the sequence from exercising to recovery to postrecovery.

Of note during `setup()` is the `MP3player.SetVolume` function. If you find the volume too loud for your headphones or amplified speaker system, try calling the function with (15,15) or even (25,25).

In the main loop, `startPin` is repeatedly read. If the momentary contact start switch is depressed, `index` is incremented, which changes the play sequence. The exception is when the current file being played is the recovery sequence. It doesn't make physiologic sense to move from the middle of a recovery sequence to postexercise heart rate, so the system is returned to the exercise heart rate. For this reason, there is no testing for the condition (`index` ==1) in the `readStartPin()` function.

Similarly, when the recovery sequence track is finished playing, the system automatically progresses to the appropriate postrecovery rate. This is accomplished in the main loop, where `index` is set to 2 if the current value of `index` is 1.

Each of the track-assignment routines reads the Hall effect sensors and assigns tracks depending on the location of the stethoscope/magnet. Both the `RecoveryRate()` and `PostRecoveryRate()` functions assign tracks based on the position of S2, the fitness switch, which is read into the port assigned to `fitPin`. If a magnet is not detected by any of the sensors, the default track, the aortic track, is played. Because the recovery sound file is two minutes in duration, you won't get a second chance to position the stethoscope/magnet once the track is selected.

Operation

First confirm that the microSD memory card contains all the MP3 files from www.mhprofessional.com/Androids. Because the program reads files sequentially, if you skip files, the MP3 on the card files will play out of sequence, if at all.

With the microSD memory card loaded in the MP3 player shield, upload the program from Listing 8-3. Within a few seconds, you should hear the heart sounds in the four different chest locations, all beating at the exercise rate of 150 beats/min. Set the fitness button, S2, to either closed (fit) or open (unfit). Next, with one hand, place the stethoscope/magnet over one of the four areas of the heart. With your free hand, touch momentary contact switch S1. Listen to the recovery heart rate over the next two minutes. When the recovery track is finished playing, the appropriate postrecovery track will continue looping until you touch S1 again to return the system to exercise mode.

Modifications

The most obvious modification is to integrate this heart rate recovery simulator with the pulmonary sound simulator, as we did in Chapter 6. You'd need pulmonary sound recordings that parallel the recovery for fit and unfit subjects. In addition to recovery-rate differences, you could consider adding wheezes and coughs to the unfit recovery and postrecovery lung sounds. This would suit both subjects who are unfit from a cardiovascular perspective and those who suffer from exercise-induced asthma.

A more challenging modification is to support changes in chest location during playback of the recovery-period heart sounds. The library supports such a feature, but it's involved. You'd have to track the time as well as the location of the stethoscope/magnet and jump to the proper time within a new track when the location of the stethoscope changes.

Gremlins

Because the projects described in this chapter are extensions and modifications of previous projects, you shouldn't run into many snags. Where you may run into challenges is expanding these simple examples to more complex systems. For example, the challenge with doing anything complex with the MP3 player shield in place is the limited number of free ports available for sensors and effectors. Moving from an Uno platform to the Mega series is one solution. There are also other MP3/MIDI player shields available that may provide more free input-output (I/O) ports.

If your aim is to approach the richness of a real cardiovascular system, then you'll have to pull your Mac/PC into the game. Although we've built physiologic synthesizers with unlimited combinations of normal and abnormal sounds, if you have limited time and resources, the next-best thing is a library of sounds that resides on your hard (or solid-state) drive. Use the Arduino for monitoring stethoscope position and Processing on the Arduino input, and you can select from and play a vast library of pulmonary and heart sounds on the fly.

Search Terms

Try the following search terms for your browser:

- Myasthenia gravis
- Muscular sclerosis
- Autoimmune disease
- Muscle weakness
- Heart rate recovery
- Exercise-induced asthma
- Heart rate monitor

CHAPTER 9

Affect and Expression

W
hat is it that makes something appear lifelike? Movement and functionality help, but people tend to pick up on certain perceptual cues that instinctively tell them that something is lifelike or a creature worth relating to as opposed to a machine. The first thing that can help with this perception is *anthropomorphism*—making things in a human-like shape convey the impression of being a creature. We see this principle in action with children's toys, dolls, and teddy bears. People are also attuned to seeing and recognizing faces. In fact, newborn infants instinctively show a preference for looking at human faces. We also see it in industrial applications: the fronts of cars resemble a face with two headlights for eyes and a bumper for a mouth. Attempts to build cars with other form factors such as a single headlamp in the middle never sold with the public. Another thing that makes something seem lifelike is action. Action can be in the form of movement or words. People like things that they can interact with and respond to them dynamically. Humans are highly verbal creatures that use language extensively; therefore, something that talks is more lifelike than something that does not. To summarize, we can make robots more lifelike if they:

1. Resemble humans in shape
2. Have faces, especially with expressions
3. Do things, preferably interactive things
4. Talk

High-Fidelity Approaches

One effective way to make a realistic android is to go for a high level of fidelity and realism to create very human-like robots. David Hanson created an android company that makes highly realistic robots that he calls "Genius Machines." Dr. Hanson's goal is to combine high levels of realism with advanced interactivity, cognitive artificial-intelligence capabilities, and effective emotional expression. A typical high-fidelity robot head will be made of special skinlike materials and have up to 32 servos attached to tethers that simulate facial muscles. The result is an android that not only looks realistic but also can reshape its face dynamically to change expression. These robots can smile, grin, grimace, wink, look worried, and snarl. When these capabilities are combined with effective speech and interactivity, people start treating these androids like people. Those interested in this approach will find a visit to www.hansonrobotics.com to be useful for science-article references and videos of these high-fidelity androids in action.

The benefits of the high-fidelity approach are obvious (Figure 9-1). The tradeoff is that the amount of development work and expertise required are beyond those of most people. Necessary skills include materials science for constructing lifelike materials for tissue and knowledge of anatomic principles. Building useful expressions with meaning benefits from knowledge of facial-action coding, a field of psychological science. On top of that, engineering skills are required to construct three-dimensional models, rigs, electronics, and animatronics. Sequencing all of this in software is also a big task (Figure 9-2).

A major problem with going for the high-fidelity approach is that people are very sensitive to things that look a little bit off. This means that a high-fidelity robot face needs to look and work perfectly or people will feel that it is "creepy" and unnatural because of the *uncanny valley effect*: psychological research has shown that people will accept a robot or software avatar as realistic and personable if the fidelity is low and cartoonish. They will do the same if the fidelity is very high, but

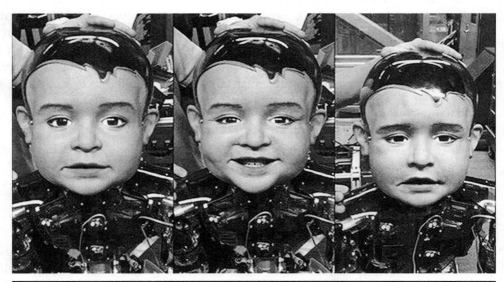

FIGURE 9-1 Three facial expressions from the Diego-San, an android child created at University of California San Diego (UCSD). (*Courtesy of David Hanson.*)

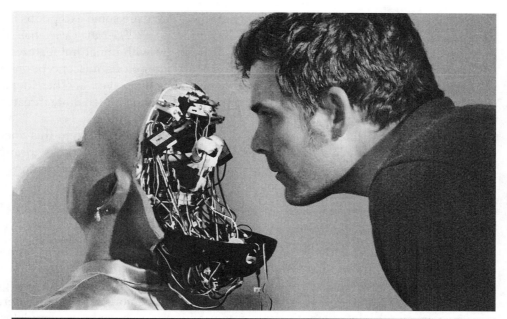

FIGURE 9-2   Dr. Hanson inspects the innards of the high-fidelity android Bina. (*Courtesy of David Hanson.*)

any parts of a high-fidelity character that don't ring true will alarm the person interacting with it. Because of this uncanny valley effect, experimenters will find a lot more room for error with low-fidelity androids.

Low-Fidelity Approaches to Androids

In this chapter, we are going to look at several low-fidelity approaches to making expressive androids that people will like and enjoy interacting with. Don't feel bad about going low fidelity, however. If you think of it, the most popular and beloved robot movie characters in the world are low-fidelity representations. C3PO, R2D2, Wall-E, the Iron Giant, those Short Circuit robots, and so on are low-fidelity robots that express themselves very well. C3PO expresses itself through humorous voice acting. R2D2 uses lights and musical sounds. Wall-E uses musical sounds and expressions that mimic the human face. All these robots are anthropomorphic (except R2D2), all are abstract in form, and all are low-fidelity representations of humans.

Now think of all the creepy robots you remember from the movies. They are often high-fidelity models such as Bishop in *Aliens*, Arnold in *The Terminator*, the *A.I.* characters (played by humans), and the *I-Robot* androids. People instinctively seem to feel threatened by robots that are too human-like. Movie screenwriters take advantage of this bias and use human actors and human-like robots as villains.

Note that these are all high-fidelity approaches. There are some exceptions to this "creep-out effect" such as the android Data in *Star Trek: The Next Generation*. Even though Data is high-fidelity, the actor was made up with unnatural features. The actor's skillful portrayal exhibited stiffness and humorous misconceptions that made him seem less threatening. The same actor also played another identical-appearing android, Lor, that acted more human-like and seemed quite threatening in comparison.

The plan for this chapter will be to integrate technologies from this book to create abstract and low-fidelity representations that use a few tricks to appear lifelike. This will include anthropomorphic design and use of expressions and voice.

Servo Smiles

One of the easiest tricks to creating an expressive robot face is to make mouth expressions with two servos and a rubber band. Most electronics hobbyists have a couple of servos lying around anyway, and the servo smile project is a great introduction to creating expressions. The approach demonstrated here generates an open mouth, closed mouth, smile, frown, and smirk. The code also simulates the appearance of chewing (Figures 9-4 through 9-9). That's a lot of expression for one rubber band!

Servo smiles are enjoyable to watch in action and make for a decent children's activity. The code for servo smiles is almost trivial. In our example, two servos are glued to a board, and a rubber band is run around posts connected to the servo armatures (Figure 9-3). A hot-glue gun is enough to stick the servos on. The only tricky part is to make sure that the servos are synchronized. You can do this by turning the servos all the way counterclockwise, unscrewing the armatures, and then reattaching them so that they are completely horizontal.

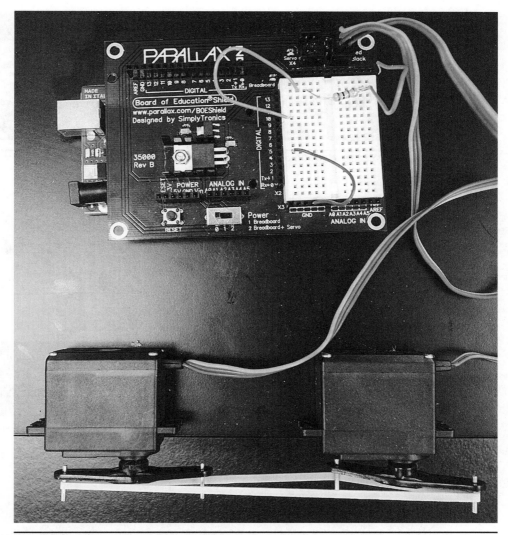

Figure 9-3 Servo Smiles using a Parallax board of education shield and two servos glued to a board. Note the use of metal posts in the armature holes—they are used to hold the rubber band onto the servos.

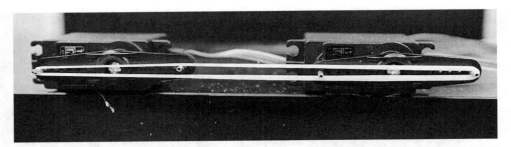

FIGURE 9-4   Servo smile, closed mouth.

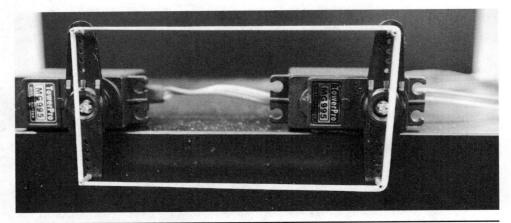

FIGURE 9-5   Servo smile, open mouth.

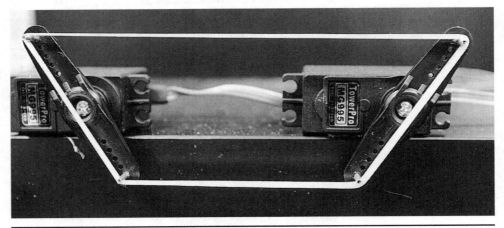

FIGURE 9-6   Servo smile, giving us a nice big smile.

FIGURE 9-7   Servo smile, big frown.

FIGURE 9-8   Servo smile, smirking.

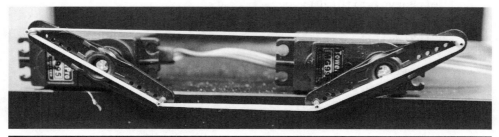

FIGURE 9-9   Servo smile, wide and subtle smile. This is also a useful open mouth stage when used for the animation of chewing. Chewing will alternate in appearance between Figure 9-9 and Figure 9-4.

LISTING 9-1   Servo smiles program code.

```
/*
Servo Smiles - A Cheap and Easy Method of Expression
Androids: Build Your Own Lifelike Robots by Bergeron and Talbot
Listing 9-1.
By Thomas Talbot
This is configured for the servo port on the Parallax board of
education shield, I/O 10 & 11
Other shield with servo ports will work just fine by changing the
I/O numbers
Use an external power supply with this code: servos draw a lot of
power.
Arduino 1.0.1 environment
 */
#include <Servo.h>

Servo leftServo;   // create servo object to control a servo
Servo rightServo;
int wait = 5000;

void setup()
{
  leftServo.attach(11);   // attaches the servo to a pin for the
                          // servo object
  rightServo.attach(10);
}

void loop()
{

    //open mouth
  leftServo.write(90);
  rightServo.write(90);
  delay(wait);

    //Smile
  leftServo.write(60);
  rightServo.write(120);
  delay(wait);

    //Frown
```

```
   leftServo.write(120);
   rightServo.write(60);
   delay(wait);

     //Mouth part open - wide smile
   leftServo.write(20);
   rightServo.write(160);
   delay(wait);

     //Smirk
   reset();
   leftServo.write(0);
   rightServo.write(100);
   delay(wait);

     //Close Mouth
   leftServo.write(0);
   rightServo.write(0);
   delay(wait);

     // chewing
   for (inti = 0; i< 20; i++)
   {
     leftServo.write(20);
     rightServo.write(160);
     delay(250);
     leftServo.write(0);
     rightServo.write(0);
     delay(250);
   }

   reset();
   delay(10000);
}

void reset()
{
   leftServo.write(90);
   rightServo.write(90);
   delay(250);
}
```

Bill of Materials

- Arduino Uno or similar
- Parallax board of education (35000) (www.parallax.com) or your favorite shield with servo ports
- 180-degree Futaba servos (2) or similar

Speech Synthesis

An interactive android needs a voice. There are two approaches you can take: recorded sounds or voice synthesis. Recorded sounds allow for a real human voice and a wide range of expression at the cost of having to prerecord everything that your android is going to say. Hardware for this approach is the same as the MP3 player–based heart sound simulator found in Chapter 6, but with the use of a louder amplifier and speaker. The biggest problem with the prerecorded approach is whether you want to add additional voice content in the future. It is challenging to get a voice recorded months or years later to sound the same as the existing vocabulary. This means that adding content usually means rerecording everything.

Voice synthesizers sound less natural but are reasonably cheap and allow your android to say just about anything in a dynamic fashion. Some microcontroller users will be familiar with the inexpensive SpeakJet chip, a phonetic voice synthesizer for which a language-to-phoneme support chip is also available. SpeakJet is easy to use and produces a very robotic monotone voice that can be useful for some applications. Language coming from a SpeakJet is sometimes hard to understand, though.

A newer synthesizer is the EMIC-2 text-to-speech module. EMIC-2, designed by Grand Idea Studio for Parallax, is specifically designed to work with microcontrollers. It has a built-in word library, so use of phonemes is not required. The speech is high quality in both English and Spanish with dynamic control of speech. The dynamic control is important because it allows for changes in voice characteristics, word emphasis, speech rate, and volume. EMIC-2 (pictured in Figure 9-10) employs an Epson speech chip and has a very simple command set. The module includes nine predefined voices, including voices of males, females, and a child. The Spanish voices are also very good. EMIC-2 also supports DECtalk, a standardized speech-synthesis format mostly used in communication devices for the handicapped. DECtalk is more complicated to interface with but allows for phonemic control and, more important, pitch. Such control can even allow your android to sing a few songs. The module works off a single serial line at 9,600 baud and needs 5 V to operate. It has an audio line out as well as a built-in amplifier that can drive a small 8-Ω speaker with a fair amount of volume.

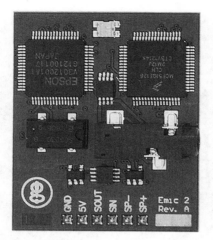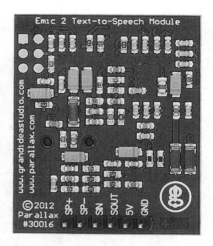

FIGURE 9-10   EMIC-2 speech module.

One of the more interesting sites for DECtalk is www.theflameofhope.co. It is a personal webpage by Pastor Snoopi Botten. He has severe cerebral palsy and uses a synthesizer to communicate. For Pastor Botten, speaking is not enough because his dream was to become a singer. Using DECtalk, he has given a number of live music performances, including a time when he sang the national anthem for a major league baseball game. Of benefit to readers, the website has a library of free DECtalk-encoded songs that can be used with EMIC-2. This book will have two examples of the DECtalk parser, one of which will be the "Star Spangled Banner." Additional information on DECtalk resources can be found in the EMIC-2 documentation and on the Web. Most of our examples will use the Epson parser because it is very easy to use.

Connecting the EMIC-2 to your microcontroller is trivial. After providing voltage, use pins 6 and 7 for software serial on an Arduino Uno. The online code sample is 9-02. If you are using the Arduino Mega, there is a bug in the software serial for the Mega at the time of this writing, so you will be better off using the Serial2 hardware serial port. The Mega has plenty of serial ports, and it will be the required microcontroller for the remaining projects in this chapter. The online code sample 9-02b is specific to the Arduino Mega. Refer to the photo (Figure 9-11) and schematic (Figure 9-12) for construction details.

FIGURE 9-11   Connecting the EMIC-2 to an Arduino Uno.

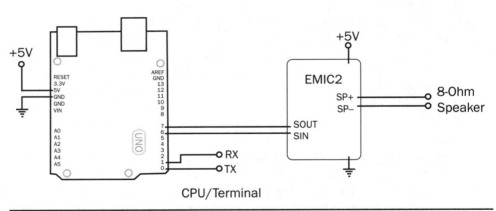

FIGURE 9-12   EMIC-2 test circuit using an Arduino Uno and board of education shield with EMIC-2 and speaker.

EMIC-2 Command Set

Sx   Convert text to speech: **x** = message (1,023 characters maximum)
Dx   Play demonstration message: **x** = 0 (speaking), 1 (singing), 2 (Spanish)
X   Stop playback (while message is playing)
Z   Pause/unpause playback (while message is playing)
Nx   Select voice: **x** = 0–8
Vx   Set audio volume (dB): **x** = –48–18
Wx   Set speaking rate (words/minute): **x** = 75–600
Lx   Select language: **x** = 0 (US English), 1 (Castilian), 2 (Latin Spanish)
Px   Select parser: **x** = 0 (DECtalk), 1 (Epson)
R   Revert to default text-to-speech settings
C   Print current text-to-speech settings
I   Print version information
H   Print list of available commands

Simple Speech Demonstration

A basic program to interface with the EMIC-2 follows. The program includes a few functions to make using the EMIC-2 more convenient. The function `emicTalk()` sends a text string for conversion to speech. All spoken statements must end with a period. This function will append the **S** speak command to the beginning of your phrase and wait for completion of speaking before returning to the caller. The function `emicCommand()` allows you to send a command to the unit. With this approach, your microcontroller will be frozen until the EMIC-2 is done speaking. The advantage of this is that your program will know when speech is completed, and it is easy to keep speech and robot actions in sequence. If you want to animate or perform actions while EMIC-2 is speaking, then you can issue commands directly to the serial port and check to see if the colon (:) confirmation code is received by your Arduino at a later time. The meat of the program is the `emicTalk("Androids. Build your own lifelike robots.")` statement. We will modify this statement in order to test out different phrasings and vocal expressions.

LISTING 9-2   EMIC-2 voice synthesis program.

```
/*
EMIC-2 Voice Synthesis
Androids: Build Your Own Lifelike Robots by Bergeron and Talbot
Listing 9-2.
By Thomas Talbot
Arduino 1.0.1 environment
 */
```

```
#define rxPin7
#define txPin6
#define CR 13                // Carriage return

// set up a new serial port
SoftwareSerialemicSerial=  SoftwareSerial(rxPin, txPin);

void setup()
{
  Serial.begin(9600);        // debug port on computer
  pinMode(rxPin, INPUT);   // set up pins
  pinMode(txPin, OUTPUT);
  emicSerial.begin(9600);
  Serial.println("Waiting for EMIC 2.....\n\n"); // \n is code for
                                            // extra line feed
  delay(4000);              // wait for EMIC 2 to complete three
                            // second initialization routing
  emicSerial.flush();
  emicSerial.write(CR);      waitForEmic();
  delay(1000);                         // Short delay
  Serial.println("Received boot from EMIC 2\n");
}

void loop()  // Main code, to run repeatedly
{
  Serial.println("Speaking....");
  emicCommand("V18");        // make volume loud
  emicTalk("Androids. Build your own lifelike robots.");
  Serial.println("\nDone speaking now.");
  for(;;) {}
}

// sends an unedited string to EMIC 2 and waits for completion
// response
voidemicCommand(String astring)
{
  emicSerial.println(astring);
  waitForEmic();
}

// adds speak command, sends string of test to EMIC 2, waits for
// completion of speech
```

```
voidemicTalk(String astring)
{
  emicSerial.write("S");
  emicSerial.println(astring);
  waitForEmic();
}

voidwaitForEmic()
{
  while (emicSerial.read() != ':');   // wait to receive ":",
                                      // indicating EMIC 2 ready to
                                      // receive commands
  emicSerial.flush();
}
```

Bill of Materials

- Arduino microcontroller
- EMIC-2 SPEECH MODULE (30016) (www. Parallax.com)
- 8-Ω, ½-W speaker

Expressive Voices with the Epson Parser

After running the preceding code sample, you got to hear a reasonably good though somewhat electronic recorded voice. If we modify the emicTalk() statement, we can change the speech output without changing any words. For example:

```
emicTalk("Androids. Build your own __expressive robots.");
```

This will emphasize the word *expressive*. Now try this:

```
emicTalk("Androids. Build your own ##expressive robots.");
```

Now the word *expressive* is whispered. Okay, for something really different, try this:

```
emicTalk("Androids. Build your own :-)3 expressive :-)0 robots");
```

In this example, the word *expressive* changes to a female voice and then reverses back to the default male voice for the last word. One big advantage of the Epson parser is the fact that the commands are all inline text. One of the important limitations, however, is that speech strings going to the EMIC-2 are limited to 1,023 characters and must end with a period. Here are more Epson parser codes that you can try out:

\ /	Decrease pitch
/ \	Increase pitch
>>	Increase speaking rate
<<	Decrease speaking rate
___	Emphasize the next word
##	Whisper the next word
:-)x	Select voice (x = 0–8)

These parser commands can also be repeated, so you can say, "I'm going to speak \ / \ / \ / very low, / \ / \ / \ right now." Try replacing your own text in the program, or experiment with connecting a terminal program directly to the EMIC-2 so that you can type commands. Another idea for using this parser is to try to associate your android's status with some sort of verbal expression. A tired or dysfunctional robot may speak slowly, whereas an excited one may speak higher and with a faster rate. While you are at it, you can use the `select voice` command to change how your android sounds. The voices codes are

0	Perfect Paul
1	Huge Harry
2	Beautiful Betty
3	Uppity Ursula
4	Doctor Dennis
5	Kit the Kid
6	Frail Frank
7	Rough Rita
8	Whispering Wendy

Expressive Voices with the DECTalk Parser

Code Listing 9-3, specifically written for the Arduino Mega, includes a `DECTalk()` command in addition to the `emicTalk()` command with which you are now already familiar. The online version of Listing 9-3 contains the full program listing with comments. Because timing issues with the module can make things not work as expected, this code goes the extra length to verify that the EMIC-2 module is ready before sending further commands. An example `DECTalk` command follows:

```
DECTalk("[:rate 200][:n0][:dv ap 90 pr 0] Rebel starship located.
Destroy all humans.");
```

This command produces a low-monotone robotic voice. The code that produces DECTalk is a fairly simple addition to our simple voice synthesis demo:

```
voidDECTalk(String astring)
{
  delay(100);
  emicCommand("P0");           // Switch to DECTalk
  Serial.print("DECTalk: ");
  Serial.println(astring);     // debug output
  Serial2.write("S");          // EMIC Speech Command
  Serial2.println(astring);
  waitForEmic();
  emicCommand("P1");           // Return to EPSON parser
}
```

The Listing 9-3 program also contains DECTalk code for the "Star Spangled Banner." It is impressive and entertaining to hear. The complexity of DECTalk becomes obvious when you see what is involved in creating a song. It requires knowledge of phonemes and sound frequencies. This is an example showing the first words, "Oh say can you see?" of the "Star Spangled Banner":

```
DECTalk("[:phone arpa speak on][:rate 100][:n0][ow<200,18>ow<200,
15>sey<400,11>kae<400,15>n yu<400,18>w siy<600,23> _<300>
```

LISTING 9-3   EMIC-2 voice synthesis for Arduino Mega, full version.

```
/*
EMIC-2 Voice Synthesis For Arduino Mega
DECTalk Demonstration
Androids: Build Your Own Lifelike Robots by Bergeron and Talbot
Listing 9-3.
By Thomas Talbot
Arduino 1.0.1 environment
 */


#define CR 13      // Carriage return

// Set up code called once on start-up
void setup()
{
  Serial2.begin(9600);    // EMIC 2 connected on pins 16 tx & 17 rx
  Serial.begin(9600);     // debug port on computer
```

```
   Serial.println("Waiting for EMIC 2.....\n\n"); // \n is code for
                                                   // extra line feed
   delay(4000);                // wait for EMIC 2 to complete three
                               // second initialization routing
   Serial2.flush();
   Serial2.write(CR);     // Send a CR in case the system is already
                          // up
   waitForEmic();
 // emicCommand("R");        // Reinitialize setting in case system
                             // resetting from warm book
   delay(1000);                              // Short delay
   Serial.println("Received boot from EMIC 2\n");
   emptyReceiveBuffer();
}


void loop()  // Main code, to run repeatedly
{

   // Speak some text
   Serial.println("Speaking....");
   emicCommand("V10");  // make volume loud if using direct speaker
                        // connection
   emicTalk("Androids. Build your own lifelike robots.");
   emicTalk("By Bryan Bergeron and Thomas Talbot.");
   emicTalk("Thank you very much for purchasing our book.");
   emicTalk("Robot voice demonstration.");
   DECTalk("[:rate 200][:n0][:dv ap 90 pr 0] Rebel starship located.
     Destroy all humans.");
   emicTalk("and now ladies and gentlemen, our national anthem.");
   DECTalk("[:phone arpa speak on][:rate 100][:n0]
   [ow<200,18>ow<200,15> sey<400,11> kae<400,15>n yu<400,18>
   w siy<600,23> _<300> bay<350,27> dhah<50,25> dao<400,23>
   nz rr<400,15>ll iy<400,17> llay<600,18>t _<300> wah<200>
   t sow<200> praw<600,27>dlliy<200,25> wiy<400,23> hxey<400,22>
   eld _<300> ae<300,20>t dhah<100,22> tway<400,23> llay<400>
   ts llae<400,18>st glliy<400,15>m iy<200,11>nx _<300>
   hxuw<300,18>z brao<100,15>d stray<400,11>ps ae<400,15>
   nd bray<400,18>t stah<600,23>rz _<300> thruw<300,27> dhah<100,25>
   peh<400,23> rrel<400,15> ah<400,17>s fay<600,18>t _<300>
   ow<200,18>r dhah<200,18> rrae<600,27>mpah<200,25>rts wiy<400,23>
   waa<600,22>cht wrr<300,20> sow<100,22> gae<400,23>llah<400>
   ent lliy<400,18> striy<400,15>miy<200,11>nx _<300>][:n0]");
```

```
    DECTalk("[:phone arpa speak on][:rate 100][:n0][ae<300,27>
    nd dhah<100> raa<400> keh<400,28>ts r eh<400,30>d glley<700>
    r _<300> dhah<100,28> baa<400,27>mz brr<400,25>stih<400,27>
    nx ih<400,28>n ey<600>r _<300> gey<400>v pruw<600,27>f
    thruw<200,25> dhah<400,23> nay<900,22>t dhae<300,20>d
    aw<100,22>rr fllae<400,23>g wah<400,15> stih<400,17>ll
    dheh<600,18>r _<300> ow<400> sey<400,23> dah<400>z
    dhae<200,23>ae<200,22>t stah<400,20>r spae<400>ngel<400>d
    bae<400,25>nrr<200,28>rr<200,27> yxeh<200,25>eh<200,23>t
    wey<800,23>ey<150,25>ey<150,23>ey<1200,22>v _<900>
    fow<300,18>rdhah<300> llae<900,23>ae<400,25>nd ah<300,27>
    v dhah<300,28> friy<1000,30>iy<1000,35> _<900> ae<300,23>
    nd dhah<300,25> hxow<1000,27>m _<600> ah<300,28>v dhah<1000,25>
    brrey<1500,23>v _<900>][:n0]");

    Serial.println("\nDone speaking.");
    for(;;) // do nothing
    {
    }
}


void DECTalk(String astring)
{
 delay(100);
 emicCommand("P0");     // Switch to DECTalk
 Serial.print("DECTalk: ");
 Serial.println(astring);  // debug output
 Serial2.write("S");       // EMIC Speech Command
 Serial2.println(astring);
 waitForEmic();
 emicCommand("P1");     // Return to EPSON parser
}


// sends an unedited string to EMIC 2 and waits for completion
// response
void emicCommand(String astring)
{
 Serial2.println(astring);
 waitForEmic();
}


// adds speak command, sends string of test to EMIC 2, waits for
```

```
// completion of speech
void emicTalk(String astring)
{
 Serial.print("EPSON: ");
 Serial.println(astring);    // debug output
 Serial2.write("S");         // EMIC Speech Command
 Serial2.println(astring);
 waitForEmic();
}

void waitForEmic()
{
  while (Serial2.read() != ':'); // wait to receive ":", indicating
                                 // EMIC 2 ready to receive
                                 // commands
  delay(100);                    // this seems to make things more
                                 // reliable
  emptyReceiveBuffer();
}

void emptyReceiveBuffer()
{
  while (Serial2.read() != -1 ); // wait for receive buffer to
                                 // empty
}
```

Building an Expressive Android

As much fun as it is to construct projects on a workbench, eventually we have to get around to building a robot. Deciding on how your android looks and what functions it can perform is a very personal decision. For this chapter, we will take advantage of an existing robot that happens to be around the office. For some reason, we happen to have an early prototype of the Vecna BEAR robot sitting around (Figure 9-13), a benefit of working at a U.S. Defense Department laboratory. BEAR originally stood for "Battlefield Extraction and Assist Robot." Current versions are highly mobile and can balance on the tips of their tracks, climb stairs, rescue soldiers under fire, conduct reconnaissance, and defuse bombs. A search of "Vecna BEAR" on the Internet will yield all sorts of interesting videos. A major limitation of our early prototype is the fact that it has a goofy bear-shaped Styrofoam head that is not functional. This robot is screaming out to be made into a more lifelike android.

FIGURE 9-13   BEAR on the prowl. Here is a more advanced version of the BEAR rescuing a solder casualty manikin during tests.

Our solution to making this robot more lifelike will be to add a different head that incorporates elements from this book. It will rely heavily on Chapter 4 and its light-related projects. We will start with an abstract approach using a rectangular cabinet and light-emitting diode (LED) matrix eyes derived from the "Mark 54 Pupil Exam Simulator" from Chapter 4 plus object-tracking and speech-synthesis capabilities. We will put everything together into an instrument cabinet, paint it, and prepare it for mounting. See Figures 9-14 through 9-20 for a step-by-step view of the android head's construction. We will then give the new android a unique name, Tatrick.

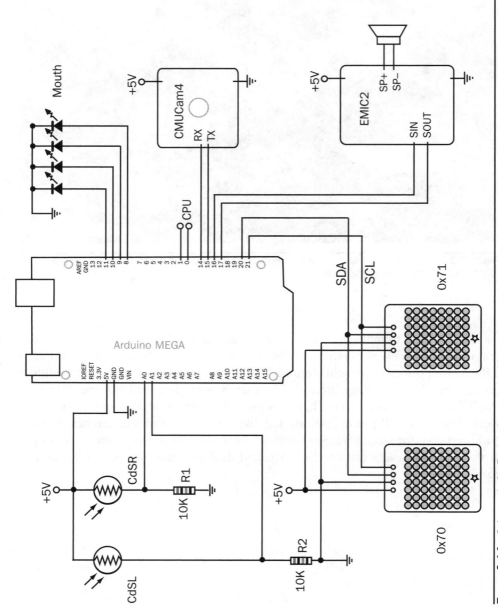

Figure 9-14 Schematic for Tatrick the android. Many of these elements will appear familiar from elsewhere in this book. The circuit starts with an Arduino Mega and adds cadmium-sulfide (CdS) photoresistors, the CMUcam4 for object tracking, LEDs for the mouth, and the EMIC-2 voice synthesizer with a speaker.

FIGURE 9-15   This instrument cabinet holds all components from the schematic. Visible are the Arduino Mega, EMIC-2, 8-Ω speaker, mouth LEDs, and wiring. Note the center hole for mounting.

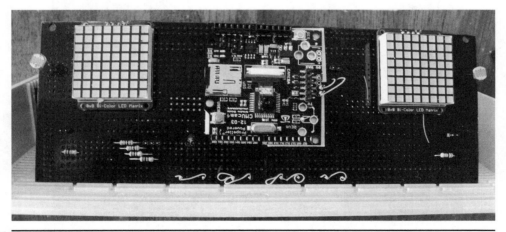

FIGURE 9-16   Front circuit board for face displays and instruments. Components are placed on two black-painted perf boards. Visible components include two 8 × 8 LED matrix displays for eyes, the CMUcam4 at the nose location, and two CdS photoresistors.

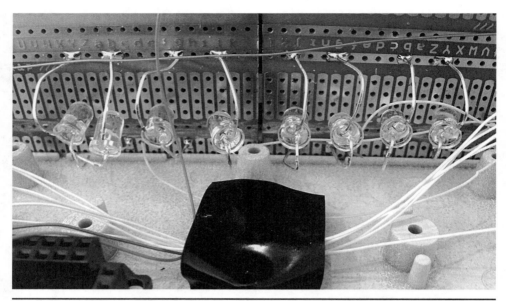

FIGURE 9-17 Close-up of front-facing LEDs. These are superbright RGB LEDs with a built-in color sequencer. In fact, they are so bright that they could be distracting, so instead they were mounted on the back of the printer circuit board. In an unusual design choice, they are mounted upside-down so that they will be front-facing and shine through the perf board holes.

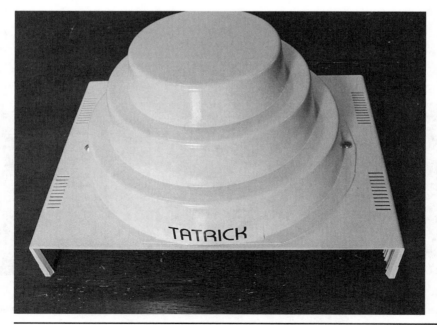

FIGURE 9-18 Top view of android head. The top of the cabinet is mounted with a plastic hat and painted yellow. Note the 1-in. hole in the center of this because the BEAR has a 0.75-in. pipe for mounting the head.

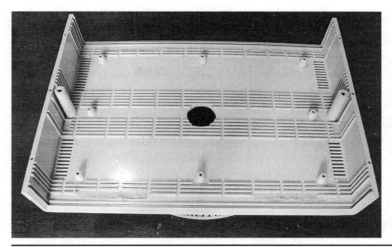

FIGURE 9-19   The completed Tatrick android head. Visible are the apertures for the eyes, mouth, photocells, and CMUcam4. The front plate is Plexiglas. Openings were covered with tape or stickers, and black spray paint was applied. On removal of the tape/stickers, clear openings remained. The neck is a section of polyvinyl chloride (PVC) pipe.

FIGURE 9-20   Tatrick joins the BEAR. The Tatrick head replaces the static BEAR head to create a more lifelike android. Tatrick will serve as a welcome guide and an interactive informational exhibit at the U.S. Army Telemedicine and Advanced Technology Research Center innovation lab at Fort Detrick, Maryland.

Bill of Materials for Tatrick

- Arduino Mega
- CdS photoresistors (2) (276-1657) (Radio Shack)
- Adafruit 8 × 8 bicolor LED matrix displays with I2C backpack (2) (www.adafruit.com)
- CMUcam4 (available at www.robotshop.com or www.sparkfun.com)
- EMIC-2 speech module (30016) (www.parallax.com)
- LEDs for mouth (RGB scintillating LEDS used here)
- Instrument case, Jameco ValuePro 10 × 7.5 × 3.10 in. (18877) (www.jameco.com)
- Plastic hat, trimmed to fit (search "DEVO whip it hat")
- Two large perf boards
- Plexiglas (for front panel)
- Resistors
- PVC pipe (neck)
- Paint and hardware

Tatrick Expresses Itself

Now that we have an integrated capability by building Tatrick or your own unique version, let's try putting it through its paces. We will take a common code base and modify it to give the same hardware a different appearance and use different voice characteristics to show variation in affect and the android's ability to express itself. This will be a very entertaining exercise as well as something that calls out for you to customize. We will start with the base code and then progress through each segment of demonstration code by section.

LISTING 9-4   Tatrick the android—self-expression example.

```
/*
TATRICK THE ANDROID — Self-Expression Example
Androids: Build Your Own Lifelike Robots by Bergeron and Talbot
Listing 9-4. See www.mhprofessional.com/Androids for fully
documented and complete code sample.
By Thomas Talbot
Arduino 1.0.1 environment

*/

// setup up eyes
#include <Wire.h>
#include "Adafruit_LEDBackpack.h"
```

```
#include "Adafruit_GFX.h"
Adafruit_BicolorMatrixmatrixRight = Adafruit_BicolorMatrix();
Adafruit_BicolorMatrixmatrixLeft = Adafruit_BicolorMatrix();
#define CR 13        // Carriage return

// Bitmaps stored in FLASH memory
static uint8_t __attribute__ ((progmem)) leftBrow[]={30, 120, 0, 0,
   0, 0, 0, 0, };
static uint8_t __attribute__ ((progmem)) rightBrow[]={120, 30, 0,
   0, 0, 0, 0, 0, };
static uint8_t __attribute__ ((progmem)) littleEye[]={0, 0, 0, 60,
   126, 102, 102, 60};
static uint8_t __attribute__ ((progmem)) pupil[]={0, 0, 0, 0, 0,
   24, 24, 0};
static uint8_t __attribute__ ((progmem)) crossedLeft[]={0, 0, 248,
   248, 248, 248, 248, 248};
static uint8_t __attribute__ ((progmem)) crossedRight[]={0, 0, 31,
   31, 31, 31, 31, 31};
static uint8_t __attribute__ ((progmem)) sageLeft[]={1, 2, 4, 8,
   16, 32, 64, 128};
static uint8_t __attribute__ ((progmem)) sageRight[]={128, 64, 32,
   16, 8, 4, 2, 1};
static uint8_t __attribute__ ((progmem)) eyeExam[]={126, 255, 231,
   195, 195, 231, 255, 126};
static uint8_t __attribute__ ((progmem)) knockOut[]={0, 130, 68,
   40, 16, 40, 68, 130};

intwaitTime = 4000;  // amount of time between demonstrations

void setup()  // Set up code called once on start-up
{
  Serial2.begin(9600);       // EMIC 2 connected on pins 16 tx& 17 rx
    //Prepare eyes and mouth
  matrixRight.begin(0x70); // The default I2C address is 0x70
  matrixLeft.begin(0x71);  // Left eye matrix set to 0x71 by
                           // bridging A0 jumper on back
  pinMode(8, OUTPUT);
  pinMode(9, OUTPUT);
  pinMode(10, OUTPUT);
  pinMode(11, OUTPUT);
  mouthOff();
  matrixRight.setRotation(3);
```

```
      matrixLeft.setRotation(3);
        // Color values are LED_GREEN, LED_YELLOW, LED_RED
      delay(3000);                 // wait for EMIC 2 to complete three
                                   // second initialization routing
      Serial2.write(CR);           // Send a CR in case the system is
                                   // already up
      waitForEmic();
      delay(1000);                 // Short delay
      emptyReceiveBuffer();
    }

    void loop()                    // Main code, to run repeatedly
    {
      emicCommand("V18");          // make volume loud
      demoOne();
      demoTwo();
      demoThree();
      demoFour();
      demoFive();
      demoSix();
      mouthOn();
      for(;;) // do nothing
      {
      }
    }

    voidDECTalk(String astring)
    {
      delay(100);
      mouthOn();
      emicCommand("P0");          // Switch to DECTalk
      Serial2.write("S");         // EMIC Speech Command
      Serial2.println(astring);
      waitForEmic();
      emicCommand("P1");          // Return to EPSON parser
      mouthOff();
    }

    // sends an unedited string to EMIC 2 and waits for completion
    // response
    voidemicCommand(String astring)
    {
      Serial2.println(astring);
      waitForEmic();
```

```
}

// adds speak command, sends string of test to EMIC 2, waits for
// completion of speech
voidemicTalk(String astring)
{
  mouthOn();
  Serial2.write("S");        // EMIC Speech Command
  Serial2.println(astring);
  waitForEmic();
  mouthOff();
}

voidwaitForEmic()
{
  while (Serial2.read() != ':'); // wait to receive ":", indicating
                                 // EMIC 2 ready to receive
                                 // commands
  delay(100);                    // this seems to make things more
                                 // reliable
  emptyReceiveBuffer();
}

voidemptyReceiveBuffer()
{
  while (Serial2.read() != -1 ); // wait for receive buffer to
                                 // empty
}
```

Demo 1: Normal Greeting with Multicolor Eyes

This demonstration has multicolored eyes (Figure 9-21) and a neutral male voice affect. It writes multiple bitmaps onto the display one color at a time. The image includes eyebrows.

FIGURE 9-21   Colored eyes.

Code Snippet

```
voiddemoOne()
{
  matrixRight.clear();
  matrixRight.drawBitmap(0, 0, littleEye, 8, 8, LED_GREEN);
  matrixRight.drawBitmap(0,0, rightBrow, 8, 8, LED_RED);
  matrixRight.writeDisplay();
  matrixLeft.clear();
  matrixLeft.drawBitmap(0, 0, littleEye, 8, 8, LED_GREEN);
  matrixLeft.drawBitmap(0,0, leftBrow, 8, 8, LED_RED);
  matrixLeft.writeDisplay();
  emicTalk("Greetings.  My name is tatrick.  I am an expressive
            android.");
  delay(waitTime);
}
```

Demo 2: Evil Robot

The next demo takes the same bitmaps as demo 1 but substitutes the green eyes with red brows for red eyes with yellow brows. This more intimidating appearance is augmented with a DECTalk-enabled flat, deep robot voice.

Code Snippet

```
voiddemoTwo()
{
  matrixRight.clear();
  matrixRight.drawBitmap(0, 0, littleEye, 8, 8, LED_RED);
  matrixRight.drawBitmap(0,0, rightBrow, 8, 8, LED_YELLOW);
  matrixRight.writeDisplay();
  matrixLeft.clear();
  matrixLeft.drawBitmap(0, 0, littleEye, 8, 8, LED_RED);
  matrixLeft.drawBitmap(0,0, leftBrow, 8, 8, LED_YELLOW);
  matrixLeft.writeDisplay();
  DECTalk("[:rate 200][:n0][:dv ap 90 pr 0] I see pesky humans.
    Prepare for violent robot uprising.");
  delay(waitTime);
}
```

Demo 3: That Confused Kid

This demonstration employs simplified eyes that look goofy with a child's voice (Figure 9-22). The child's voice is selected with :-)5. At first, the child speaks quickly to get your attention (code >>): "Hey, you there." Finally, the word *glasses* is emphasized with the __ prefix. The only flaw of this expression is that the child's voice is often too soft, especially compared to the other more boisterous EMIC-2 voices.

FIGURE 9-22   These basic block shapes face inward to give an awkward, crossed-eyes appearance.

Code Snippet

```
voiddemoThree()
{
  matrixRight.clear();
  matrixRight.drawBitmap(0, 0, crossedRight, 8, 8, LED_YELLOW);
  matrixRight.writeDisplay();
  matrixLeft.clear();
  matrixLeft.drawBitmap(0, 0, crossedLeft, 8, 8, LED_YELLOW);
  matrixLeft.writeDisplay();
  emicTalk(":-)5 >>Hey you there, <<can you help me find my __
    glasses? :-)0.");
  delay(waitTime);
}
```

Demo 4: The Drunken Robot

FIGURE 9-23   The red X eyes are a classic indicator commonly used in cartoons and comics to indicate an incapacitated person. The deceleration code <<, followed by the one word "Hiccup" makes for a convincing inebriated robot.

Code Snippet

```
voiddemoFour()
{
  matrixRight.clear();
  matrixRight.drawBitmap(0, 0, knockOut, 8, 8, LED_RED);
  matrixRight.writeDisplay();
  matrixLeft.clear();
  matrixLeft.drawBitmap(0, 0, knockOut, 8, 8, LED_RED);
  matrixLeft.writeDisplay();
  emicTalk("Someone must have spiked my hydraulic
    <<fluid.  Hiccup");
  delay(waitTime);
}
```

Demo 5: The Wise Sage

The use of simple diagonal lines (Figure 9-24) for eyes is humorous and gives the impression of talking with a closed eye elder in a state of reflection. The voice, code ":-)2", is female and is slowed down with code "<<<<" to make her sound deliberate. The seriousness of the speech delivery is offset by the humorous content.

Code Snippet

```
voiddemoFive()
{
  matrixRight.clear();
```

FIGURE 9-24 Sage eyes consist of diagonal green lines.

```
matrixRight.drawBitmap(0,0, sageRight, 8, 8, LED_GREEN);
matrixRight.writeDisplay();
matrixLeft.clear();
matrixLeft.drawBitmap(0,0, sageLeft, 8, 8, LED_GREEN);
matrixLeft.writeDisplay();
emicTalk(":-)2 <<<<Confucius say man who lie on floor cannot fall
  off :-)0.");
delay(waitTime);
}
```

Demo 6: The Patient with Patience

The final demonstration is concise (Figure 9-25). A male voice, code :-)1, is used without inflection. Round green eyes are included. When the demonstration ends, the calling procedure turns the mouth back on, and the android remains in this state until turned off. This would be a good segue into transitioning to an application such as the eye exam, or using hardware from the sound chapter, or even demonstrating the circulatory system you may have built.

FIGURE 9-25 Nice green quick eyes from the Chapter 4 pupil exam project.

Code Snippet

```
voiddemoSix()
{
  matrixRight.clear();
  matrixRight.drawBitmap(0,0, eyeExam, 8, 8, LED_GREEN);
  matrixRight.writeDisplay();
  matrixLeft.clear();
  matrixLeft.drawBitmap(0,0, eyeExam, 8, 8, LED_GREEN);
  matrixLeft.writeDisplay();
  emicTalk(":-)1 I am ready for my eye examination. :-)0.");
  delay(waitTime);
}
```

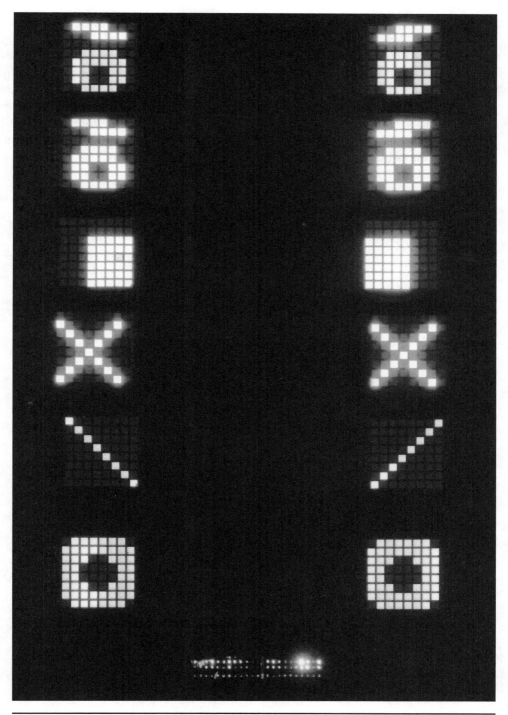

Figure 9-26 A variety of expressions from the android are shown in this composite photograph.

Summary

This concludes the affect and expression chapter as well as this book. We hope that you will find the principles of this book to be of value in making your androids seem more lifelike and approachable by people. Many of the projects in this book are based on well-known physiology in an attempt to add authenticity to your android's behaviors and actions. Likely applications include interactive exhibits; medical simulators; science, technology, engineering, and math (STEM) education; toys; robots that work with people; or experimenting for fun.

Resources

Arduino-Compatible Microcontrollers

- Sparkfun Electronics, www.sparkfun.com
- Pololu Robotics and Electronics, www.pololu.com
- Adafruit, www.adafruit.com
- Jameco Robot Store, www.jameco.com
- Maker Shed, www.makershed.com

Prototyping Supplies

- Sparkfun Electronics, www.sparkfun.com
- Seeed Open Hardware Facilitator, www.seeedstudio.com
- Pololu Robotics and Electronics, www.pololu.com
- Adafruit, www.adafruit.com
- Parallax, www.parallax.com
- Jameco Robot Store, www.jameco.com
- Maker Shed, www.makershed.com
- RobotShop, www.robotshop.com
- Digikey, www.digikey.com

Hardware

- McMaster-Carr, www.mcmaster.com
- Jameco Robot Store, www.jameco.com
- Digikey, www.digikey.com

Displays

- Adafruit, www.adafruit.com
- Jameco Robot Store, www.jameco.com
- Maker Shed, www.makershed.com
- Trossen Robotics, www.trossenrobotics.com
- RobotShop, www.robotshop.com
- Digikey, www.digikey.com

Sensors

- Sparkfun Electronics, www.sparkfun.com
- Seeed Open Hardware Facilitator, www.seeedstudio.com
- Pololu Robotics and Electronics, www.pololu.com
- Adafruit, www.adafruit.com
- Parallax, www.parallax.com
- Jameco Robot Store, www.jameco.com
- Maker Shed, www.makershed.com
- Trossen Robotics, www.trossenrobotics.com
- Digikey, www.digikey.com

Servos

- Adafruit, www.adafruit.com
- Jameco Robot Store, www.jameco.com
- Maker Shed, www.makershed.com
- RobotShop, www.robotshop.com
- Trossen Robotics, www.trossenrobotics.com
- CrustCrawler, www.crustcrawler.com

Turrets

- Trossen Robotics, www.trossenrobotics.com
- CrustCrawler, www.crustcrawler.com
- RobotShop, www.robotshop.com

Robot Arms

- Lynxmotion, www.lynxmotion.com
- Jameco Robot Store, www.jameco.com
- Maker Shed, www.makershed.com
- Trossen Robotics, www.trossenrobotics.com

- CrustCrawler, www.crustcrawler.com
- RobotShop, www.robotshop.com

Shields

- Sparkfun Electronics, www.sparkfun.com
- Maker Shed, www.makershed.com
- Parallax, www.parallax.com
- Pololu Robotics and Electronics, www.pololu.com
- RobotShop, www.robotshop.com

Arduino Technical Resources

- Tronixstuff, Tronixstuff.wordpress.com
- Arduino Playground, www.arduino.cc
- *Nuts & Volts Magazine*, www.nutsvolts.com
- JeremyBlum.com
- CMUcam website, www.cmucam.org

Combat Lifesaving and High-Fidelity Robots

- Vecna, Inc., www.vecna.com
- Hanson Robotics, www.hansonrobotics.com

Index